INTRATERRESTRIALS

KAREN G. LLOYD

INTRA
TERRESTRIALS

DISCOVERING THE STRANGEST
LIFE ON EARTH

PRINCETON UNIVERSITY PRESS
PRINCETON AND OXFORD

Copyright © 2025 by Karen G. Lloyd

Princeton University Press is committed to the protection of copyright and the intellectual property our authors entrust to us. Copyright promotes the progress and integrity of knowledge created by humans. Thank you for supporting free speech and the global exchange of ideas by purchasing an authorized edition of this book. If you wish to reproduce or distribute any part of it in any form, please obtain permission.

Requests for permission to reproduce material from this work should be sent to permissions@press.princeton.edu

Published by Princeton University Press
41 William Street, Princeton, New Jersey 08540
99 Banbury Road, Oxford OX2 6JX

press.princeton.edu

All Rights Reserved

ISBN 9780691236117
ISBN (e-book) 9780691236124

British Library Cataloging-in-Publication Data is available

Editorial: Alison Kalett and Hallie Schaeffer
Production Editorial: Natalie Baan
Text and Jacket Design: Heather Hansen
Production: Jacquie Poirier
Publicity: Kate Farquhar-Thomson and Matthew Taylor
Copyeditor: Dana Henricks

Jacket image: *Expressions in the Desert* by Landsat 8 / Earth Resources Observation and Science (EROS) Center / NASA

This book has been composed in Arno with Koulen

Printed in the United States of America

10 9 8 7 6 5 4 3 2 1

CONTENTS

INTRATERRESTRIALS

INTRODUCTION

AS I awoke the morning of what was *supposed* to be my first deep-sea dive, I prayed to no one in particular, "Oh please, don't let my legs be slamming against my mattress." If they were, that would mean the seas were too rough to launch the submersible from the research ship where we were living and working for three weeks, hundreds of miles offshore in the Gulf of Mexico. And I wouldn't get a second shot at it. As my brain gradually regained consciousness, I began to smell bacon wafting from the galley below, I heard the whirr of air conditioning, and, miraculously, I came to realize that my bunk was as still as if I were on land. The water outside was like glass. Perfect conditions.

I leaped out of my bunk and began to gather my diving clothes. When it comes to submersibles, fashion is dictated by engineering and oceanography. The bottom of the ocean is universally cold, no matter where you are on Earth, and our submersible, the *Johnson-Sea-Link II*, would not offer me much thermal protection. Though the divers in the front of the vessel sit in a beautiful clear Plexiglass ball, which insulates it from the cold and gives a splendid half-sphere view of the ocean, this cozy throne only holds one scientist and one pilot. Being one of the lower-ranking scientists (I was a PhD student at the time), I would not get to travel in such luxury. Instead, I'd be stashed in "the coffin," located at the back of the submersible. The coffin is a rectangular box where a second scientist (that would be me, for this dive) and a second pilot lay side by side, with no room for either to sit up straight. They communicate with the folks in the Plexiglass ball via an internal headset, but otherwise the ball and the coffin are separate. Since the coffin is made of metal, I would get to experience the authentic temperature of the deep sea without insulation. And, as the name suggests, this can be a deadly cold: the first version of the *Sea-Link* snagged on the bottom of the ocean, where it stayed for more than a day. The two men in the back of the submersible, one of whom was the son of Edward Link, the submersible's designer, died from cold exposure and carbon dioxide accumulation. Their tragedy led to safety improvements that now make such accidents far less likely.

So I started dressing as if I were going outside in winter. There's something incongruous about pulling on a wool sweater to get into water; a wet suit would have felt more natural. But putting on a wet suit would've been silly. The only scenario in which I'd touch seawater half a mile deep in the ocean would be if the hull breached. And if the hull breached, I wouldn't live long enough to know it had happened.

Properly attired, I joined the crew at the submersible on the back deck of the ship. Frank (the extra pilot) and I (the extra scientist) climbed into the coffin and sealed ourselves in. From our confined space, we had no ability to drive, steer, or gather samples. We were just there to observe, advise—and perhaps save everyone's lives: Frank explained to me that if the three other people were "incapacitated," I should follow a protocol to drop external weights and the submersible would bob to the surface like a beach ball. I paid close attention. I didn't want to be the graduate student who killed everyone at the seafloor because she couldn't remember which switches to flip.

A moment later, I experienced the sensation of being lifted off the deck of the ship by the powerful A-frame winch and gently nestled into the ocean, where the submersible started bobbing fitfully. After the final checks, the main pilot released air ballast and we began our descent by freefall through the ocean. Frank, who was lying head-to-toe with me, announced that he was going to sleep and wished me a nice dive. I couldn't imagine sleeping. I was wired.

The thing that you should know about the oceans is that they are not empty. Sure, there are fish and whales and sea turtles—but I'm referring to the fact that every *inch* of the ocean is packed with invertebrates and other bits of floating goo. When you're free-falling through the ocean, all this schmutz bioluminesces—it glows—when your submersible hits it. The coffin had one port-hole on each side. The one on my side was in between my shoulder and my chin—perfect for viewing the glittering ocean as long as I ignored the growing crick in my neck. Sparkly lights shot by, with the occasional crescendo of blue, red, and purple zipping through a long, segmented body. I could have hung out in the dark pelagic zone for hours, happy as a clam, just staring at the beauty.

The main pilot soon slowed to neutral buoyancy to prevent us from slamming into the seafloor. I had finally arrived at the place I had been researching for six years but had never actually visited. It was desolate. Somehow this made me like it even more. Two-thirds of the Earth is covered by oceans, and yet the seafloor remains largely unseen by humans. Chances are slim that anyone else will ever visit the exact area that I found myself in that day. It felt like I was looking at the truth.

While the folks up front discussed how to travel from where we landed to where we wanted to be, I was surprised to discover that there was more going on at the seafloor than I had antici-pated. I watched as a deep-sea crab decided to go to war with us: it held up tiny menacing claws and stood stock-still, ready to destroy us. A bright purple jellyfish floated by, and long, slith-ery fish scurried along the seafloor, searching for invertebrates to munch. But my favorite animals were the holothurians, or sea cucumbers. These creatures are passive, hollow tubes, roughly the size of two bananas laid end to end, and just as charismatic. All they do, day in and day out, is suck mud through one end of their bodies and push it out the other. They wipe the nutritious bits of organic matter from the mud and then deposit pristine beach sand out of their butts.

While I was admiring these deep-sea Roombas, the other crew members were carefully navigating the landscape. At the bottom of the sea, there are two main challenges when it comes to finding one's way around. First, because the ocean drowns satellite signals as effectively as it drowns people, there is no GPS. Instead, we improvise our own X-Y grid system by send-ing acoustic "pings" between the submersible and the ship. These pings tell us our depth and angle from the ship, and the Pythagorean theorem does the rest. Second, the lights on the submersible are no match for the darkness of the sea, so it's impossible to view the whole landscape. We'd only see an un-derwater mountain if we were about to run into it. Sonar can help avoid major catastrophes, but there's no solution for the problem of being just a few meters away from the desired site and not knowing it. We used the X-Y grid to make our way to the general area, but then we had to nose around like a wobbly beetle until we found the spot.

Our target that day was a cold methane seep, where ancient, deeply buried methane burbles up to the seafloor through cracks caused by movements of tectonic plates or geological scouring.

In our normal lives, methane is the natural gas we use to heat our houses, but at the seafloor, it's "manna from heaven" (or perhaps, more accurately, "manna from hell"). Methane is a highly energetic food in what is otherwise a desert, so life crams around it like antelopes at a savanna watering hole. The animals here don't eat the methane, but they eat the microbes (tiny, single-celled organisms that include the bacteria) that do. So, although you can't see methane in water unless there's so much of it that it forms bubbles, you can tell when you're nearing a methane seep because they are jam-packed with clams, mussels, crabs, shrimp, fish, sea anemones, and creepy, otherworldly worms.

As I lay in the submersible, craning my neck to watch the seafloor go by my awkward porthole, I started to see bits of broken shells and eventually whole, live mussels packed tight and sticking out of the muck. Crabs were crawling across the mussels, picking off filamentous tendrils of bacteria and invertebrates to eat. Soon, every place that wasn't covered in mussels was carpeted with bright white mats of Beggiatoa—bacteria that transform stinky sulfide into pearls of pure elemental sulfur. We had arrived at the methane seep! We stopped the submersible and began our work.

I had spent the previous evening hose-clamping metal T-shaped handles to cylindrical plastic core tubes to make them compatible with the robotic grippers on the front of the submersible. To keep the cores from floating away while we descended, we strapped them onto milk crates on the front of the submersible with rubber bands, which the robotic arm was able to break when it picked up the tube. Low-tech workarounds like these are a mainstay of scientific exploration—when you're doing something that very few people do, you can't just buy the equipment you need, prefabricated, from a store shelf. The main pilot used the submersible's robotic arm to pick up the core tubes, punch cores of dark black mud, and retrieve them back into the milk crates.

While the work was going on, I lay in my little coffin, occasionally conferring with the main scientist about where we should take cores. After about eight hours, the main pilot dropped the weights, and we started our half-hour ascent.

The reason I was so eager to take this perilous journey, and the reason I immediately began plotting ways to return, is that I was searching for the answer to a question that had been gnawing at me for a long while: *Are there life-forms hiding inside Earth that are so strange that they change our conception of life itself?* Let's explore this question, shall we?

The Intraterrestrials

For much of my life, I have been tracking the strange types of microbes that live at the bottom of the oceans, inside volcanoes, and deep within the Arctic permafrost, in an attempt to answer this great driving question about hidden life on Earth. From my efforts and those of other scientists like me, we have learned that life can even exist kilometers under the seafloor, way deeper than you can get to by submersible. In fact, we have not yet encountered a depth at which life ceases to exist. And most of this subsurface life is entirely unlike anything we find at the surface: against all odds, it seems that Earth's subsurface may be a nice place to live, as long as you aren't too attached to multicellularity or oxygen. Luckily for us humans, these tiny life-forms promise to unlock some of the most important mysteries of life: they might tell us how life first developed on this planet, change our basic assumptions about the rules of life, even upend our understanding of what it means to be alive. As a bonus, they might save us from our self-destructive tendencies by helping alleviate the effects of climate change.

The biogeochemist Karsten Pedersen coined the term "intraterrestrials" to describe this abundant life within Earth's crust.[1] "Intra" means inside, and "terrestrial" means "of Earth," so intraterrestrial literally means an inhabitant inside the Earth. I like this word because it mirrors the term "extraterrestrial," which conveys some sense of the alien nature of these new life-forms. Now, I don't want to diminish the importance of finding new *animals*—new species of monkeys are thrilling. But the major categories of visible life on Earth are pretty much settled. The discoveries we're making within the Earth's crust are like finding the existence of *all* animals, many tens of times over, based on the evolutionary novelty of these organisms relative to previously known life. This ongoing discovery, which started in the late 1980s, is gradually revealing that we have been missing major branches on the tree of life.[2]

Part of the reason these creatures are so different from previously known life is that, although we share a planet with them, we inhabit vastly different worlds. David Valentine, another biogeochemist, has aptly described the "microbial purgatory deep below Earth's surface," in which these single-celled organisms thrive: "Bounded from below by the inhospitable temperature of Earth's interior, intraterrestrials face a chronic limitation of food-derived energy because they are far removed from sunlight-driven productivity."[3] Once we begin to peer downward past our feet into the deep, dark recesses of Earth's crust and oceans, a new world emerges. This new world raises a host of questions, such as: Without the sun, where do these creatures get energy? Without oxygen, what do they breathe? And how long, exactly, can any organism survive in harsh environments, where pH ranges from pure acid to pure alkaline? The answers to these questions—they get energy from chemical reactions, breathe rocks, and sometimes live for thousands or perhaps millions of years—will demonstrate that our assumptions about the boundaries of life, based on our narrow experience of living in the thin green layer at Earth's surface, are often wrong.

Our journey to understand these creatures will be divided into three sections. In part I of this book, I will describe what this subsurface habitat is like, how we exhume living beings from it, and how we use DNA sequencing to "see" these microscopic beings. In part II, I will describe how these intraterrestrials have changed what we know about the evolutionary relationships among all life on Earth, how they are able to thrive in previously unthinkable environmental conditions, and how they play with thermodynamics* in ways that are totally foreign to life on Earth's surface. In part III, I will suggest that intraterrestrials skew how life interacts with time itself, give us new perspectives on life's origins, and maybe, if we play our cards right, can help us with climate change. Finally, I'll imagine what life will be like a thousand years into the future, perhaps on other planetary bodies, based on the expanded vision shown to us by the intraterrestrials.

The journey we take in this book will not merely be one of intellectual awakening. My goal is not just to explain what we've discovered about these new life-forms, but also to describe *how* we've made these discoveries. To collect samples and conduct the research described in this book, I've chased intraterrestrials to the ends of the Earth, and I want to take you with me—to Argentina's desolate altiplano; to the frozen Arctic tundra of

---

*I find the Wikipedia definition to be quite complete: "Thermodynamics is a branch of physics that deals with heat, work, and temperature, and their relation to energy, entropy, and the physical properties of matter and radiation" (https://en .wikipedia.org/wiki/Thermodynamics).

Svalbard; to the bottom of an active volcano in Costa Rica; to the muddy coastline of North Carolina; and, of course, to the bottom of the ocean.

Researching intraterrestrials involves long hours at a lab bench processing samples and at a computer analyzing data. But it also involves crawling on one's belly through spiders and bat guano, donning a mask to survive toxic fumes, and spending weeks or months in remote locations that test the limits of one's physiology. The mind-bending discoveries I describe in this book would simply not be possible without the hardships and triumphs of fieldwork. So, put on your diving clothes and climb into the coffin: we're about to embark on a journey to the depths of the Earth.

PART I

WHAT LIVES INSIDE EARTH AND HOW DO WE GET TO IT?

I

IS THERE A "HABITAT" INSIDE EARTH'S CRUST?

IT WOULD be reasonable to assume that anything living deep inside Earth's crust, far away from the reach of sunlight, would be a leftover from the surface world, marooned and condemned to die. Our lives are so sun-centric that it can be difficult to conceive of life at the bottom of the ocean, much less miles underneath the seafloor. However, even though it's remote, much of Earth's habitable space lies well below even our deepest fathoms, buried under hundreds of meters of sediments and rock. Any sad little trickle of life in this deep, dark underworld, therefore, might seem like it should be trying to clamber back up to the surface, desperate for life-giving sunlight. But anyone who's traveled to a methane seep at the bottom of the sea, like I have, knows the opposite is true: life flourishes in the dark. In this chapter I will describe how Earth's deep, dark recesses may be, counterintuitively, good habitats for life.

Life in a Strange Place

"You want to get out of the truck and run with them, don't you?" Peter Barry, a geologist at the Woods Hole Oceanographic

Institution, noticed me gazing out the window as we bounced across a desolate rock-scape in Argentina's altiplano. The only living thing in sight was a herd of vicuña, tiny llama-like animals native to the central Andes. The vicuña raced beside us on their twiggy legs as we traversed the plain between towering volcanoes.

I murmured a "yep" to Peter, idly wondering what vicuña might eat—I hadn't seen anything green for days—when Peter suddenly stopped the truck and gave me an encouraging look. Taking the hint, I burst out of the truck and started trotting alongside the herd. Predictably, the vicuña didn't share my spirit of comradery and quickly split the scene. But I kept running. In that gorgeous moment, my field of vision was filled entirely by rocks, volcanic mountains, salt crusts, and the trucks carrying my colleagues. Running was the only way I could express my joy at being in this magical place, even if my celebration was short-lived: at 4,000 meters elevation (that's about 13,100 feet), the air is so thin it makes your breath worthless. Also, I didn't travel to Argentina all the way from Tennessee to chase vicuña. I was here for the volcanoes.

If you, like me, don't live near any volcanoes, you might think they are all hollow-throated cones with boiling lava pits that occasionally flash-kill cities like Pompeii. This is true for some volcanoes, which announce their propensity for violence with steaming gas jets or by spewing lava. But most volcanoes, at least when considered on a human timescale, appear serene, masquerading as laconic piles of rocks. Don't be fooled. Beneath these rocks, Earth is in motion. And when a hot glob of magma rises to Earth's surface and meets water, conditions are ripe for a violent convection. As the magma heats up the water, the pressure increases until, eventually, the whole thing explodes. It's like placing a glass bottle of Coke in a hot oven: the heat will cause the gas inside the bottle to expand, and soon the bottle will explode, shooting shards of glass into your kitchen. The placid rocks on a pointy volcanic mountain are basically a tightly screwed on lid, waiting to explode.

Exploding piles of rocks are not particularly good at supporting life. Even the most badass life-forms will melt like a microwaved marshmallow when exposed to fresh lava. Why, then, were my South American colleagues* and I driving around Argentina looking for life inside volcanoes? Even though nothing can live in lava or magma, we wanted to find out how close life can get to the "business end" of a volcano and survive. Volcanoes are violent, but this violence has an upside: it also brings up nutrients from deep inside the Earth. We were in Argentina to hunt for any life buried deep underground that might thrive on these nutrients.

I'll spoil the suspense and divulge that we *did* find life near those volcanoes in Argentina. Lots of it. We found billions and billions of tiny microbes called bacteria and archaea, most of which are invisible to the eye, and many of which were previously unknown to science. You might be wondering why you didn't hear about our major discovery in the news. A clickbait headline like "Scientists Discover New Life-Forms Fueled by Volcanoes!" would not have been far from the truth. The reason that we didn't make waves is that discoveries of this magnitude are now commonplace. And we're not the only researchers making them. Each year, literally thousands of scientists dredge up never-before-seen microbes. And these microbes are not

---

* This work is part of the ongoing international collaboration, which, in 2019, included geologist Agostina Chiodi from Argentina and Gerdhard Jessen from Chile. The collaborations have included more folks in subsequent years.

just near volcanoes. No matter where scientists go on Earth, or *in* Earth, we are finding new forms of microbial life.

In light of these recent discoveries, we need to face the fact that we don't really have a good handle on what life is like on Earth. Both literally and metaphorically, we've only just scratched the surface. What will we find once we dig a little deeper into the subsurface?

To answer this question, we need to define at what depth, exactly, Earth's subsurface biosphere begins. There is currently no consensus about this, so I prefer the following expansive definition: the subsurface biosphere is any place below ground or below seafloor that is not regularly exposed to sunlight, but where conditions can support life. That covers a lot of real estate: from the dirt right under our feet, to subterranean caves, to oceanic sediments piled tens of kilometers deep. It includes environments that look like solid rock but are actually porous, with small cracks and fissures that allow water to move through them and life to flourish; caves and mines that bring oxygen down to meet the chemicals coming from inside Earth; deeply buried basalts and geothermal springs; and mud and sediments that have accumulated slowly and passively over millions of years. Some of the subsurface biosphere is frozen solid as permafrost and ice, where microbes eke out a living in the tiny brine veins that stay liquid below the freezing point of pure water.

Across this diversity of environments, there is one constant: life. No matter how deep we have drilled into Earth's crust, we haven't yet hit a depth where life does not exist. Earth consists of a gigantic inner core, an even larger mantle that wraps around that, and finally the thin crust at the top that we live on. The whole thing is about 7,000 km deep and the crust is only about 70 km of it. If Earth were a piece of M&M candy, the crust would be the colored paint around the top. But 70 km is extremely deep for biology and our ability to drill down to it. In fact, it's way deeper than any human has ever drilled. Despite our knowledge of Earth's mantle and core, we've never actually visited either one of them. Everything we know about them comes from minerals expelled by volcanoes or plate tectonics, or by earthquakes moving through the center of the Earth.* But the few kilometers of crust that we have drilled so far appear to be replete with life. From samples that have been pulled out of these places, scientists have been able to grow microbes and detect intact DNA, RNA, proteins, metabolites, and lipids. We can even directly measure the chemical products of the metabolisms we expect from the biomolecules we find. There's apparently a lot of fertile ground above the crust-mantle boundary, and that's where subsurface life thrives.

Food from Above and Below

Subsurface biospheres can be divided into two main types. The first depends on chemosynthesis (also called chemolithoautotrophy) rather than photosynthesis. Chemosynthesis is the process by which living organisms make biomass from pure chemical reactions. No sunlight required. The second type eats leftovers from photosynthesis that trickle down from the surface. Some environments have a mixture of the two.

In a chemosynthetic biosphere, life-giving chemicals either come up from the deeper parts of Earth, where conditions are too harsh to support life as we know it, or they are produced right next to the microbes that use them. Here's where

---

*An earthquake in the US can give scientists in China information about what sorts of materials are in the center of the Earth because the waves change speed and shape as they travel through them.

subsurface life has an advantage over the surface; they've got pressure and heat. Pressure and heat turn rocks and minerals into chemicals. Intraterrestrials use these chemicals to gain energy and build their bodies, just like plants do with sunlight at the surface. But unlike the sun, a one-trick pony that bathes Earth in photons, the life support that comes from the deep Earth is varied in its distribution, as well as in the chemicals it produces. It's as if there are millions of little low-powered suns distributed throughout Earth's crust, each one with its own tiny orbit of a subsurface ecosystem.

Up at the surface, sunlight reaches almost everything we can see. In the subsurface, chemicals have to move through a series of conduits to reach all the living cells that need them. In a person's lifespan, Earth's crust may seem stable, but over geological time, it is in constant motion. Tectonic plates slam together, jostle each other as they slide past, and pile up in heaps. This motion creates cracks that act like superhighways for these energy-rich chemicals. To picture how this works, think of pieces of modeling clay of various stiffness and density spread around the surface of a ball. If you start moving one piece of clay around the ball, all the other pieces will bunch up, deform, or tear to accommodate the motion of the piece. Now imagine that one clay piece is much heavier than an adjacent piece that it pushes into. If the ball is made of something soft and mushy, the heavy piece will sink into it, and the lighter piece will pile up on top of it. When one tectonic plate subducts under another one in this way, it's called a subduction zone, and it creates cracks and weak points in the plate that provide extra opportunities for chemical movements.

Sometimes the oceanic plate that's being pulled down is covered in seamounts—little mountains that poke up out of the seafloor. When these tectonic plates are subducted underneath a continent, they have an effect like pushing your fingers through Jell-O. They drag through the continental crust, pushing up mud, rocks, and fluid into vertical rivers called mud volcanoes. These mud volcanoes eject deep subsurface chemicals and microbes up into seawater and surface sediments.

Another activity that moves chemicals around Earth's crust is called a spreading center or rift zone, where two tectonic plates pull apart, leaving a void between them. As happens when you dig sand apart at the beach, water will rush in the resulting trench, but something else happens too. Moving two 100 km thick chunks of Earth away from each other is powerful enough to make the underlying mantle unstable. This unstable mantle breaks free and rises toward Earth's surface. As it does so, the pressure bearing down on the glob of mantle drops dramatically—not unlike when scuba divers rise to the surface too quickly and experience the bends—which lowers the mantle's melting point, causing it to liquify, which further accelerates its upward journey in a positive feedback loop. This cascading series of events causes all hell to break loose and creates a volcanic eruption. A similar mantle liquification process occurs with subducting slabs as well because the sinking oceanic plate injects the mantle with water, causing magma to ascend. This latter process, known as arc volcanoes, throws up spectacular mountain chains like the ones running through the west sides of South, Central, and North America, as well as Japan, Indonesia, and elsewhere.

Just as is true for subduction zones, volcanic eruptions leave gaping cracks and fissures in what was previously solid rock, providing pathways for these chemicals to move around. Even during times when everything is still, seawater continues to be sucked deep underground by the thermal convective cells created by the heat gradient that is always present in Earth's subsurface. Once seawater reaches the deep subsurface, it becomes superheated and loaded with metals and other chemicals. When it shoots back to the surface, it mixes with different chemicals in the seawater. At all stages of this process, life can pocket the energy spilling out of these reactions to fuel the subsurface biosphere. If you think of the Earth as an organism, rather than a habitat, these deep cracks function like veins and arteries, bringing life-giving nutrients to cells. All this void filling has spectacular effects on the seafloor—creating giant suture-like cracks running down the middle of the Pacific, Atlantic, and Indian Oceans, like the seams of a baseball.

But you don't have to rip through an entire oceanic plate to deliver energetic chemicals to the subsurface biosphere. Some places like the Gulf of Mexico and the Mediterranean Sea have deeply buried salt deposits, left over from ancient lakes that dried into salt crusts millions of years ago. These salt deposits are more buoyant than the sediments that cover them; as they force their way slowly through the sediments, they break open new conduits for deep subsurface fluids along the way.

All these processes happen on land too. But here, instead of pulling down seawater, the heat convection pulls down what geologists refer to as "meteoric water." When I first heard this phrase, I thought it meant water from meteors (awesome!), but it turns out that it just refers to old, warmed-up rain. These movements create habitats for microbes where chemicals, not sunlight, might form the base of the ecosystem. I say "might" because most of these places have yet to be explored.

So that's how life-giving chemicals move around inside of Earth's crust, but how do they get there in the first place? One massively powerful reaction that produces these chemicals occurs when minerals called olivines and peridotites mix with water under high pressure and temperature to form hydrogen and reduced iron. This process, called serpentinization, is currently one of the top candidates for fueling the first life on Earth because it provides all the chemicals needed for life, without depending on sunlight to oxidize any of them. Chemicals are also made when tectonic plates sink through subduction zones (recall the clay pieces on the ball). Here, carbon dioxide, carbon monoxide, metals, sulfur, and nitrogen are squeezed out of the minerals and dissolved into water, where they can be used by microbes. Even something as simple as the friction of rocks moving against each other during earthquakes creates life-giving chemicals. The chemolithoautotrophs that live off these chemicals, can, in turn, feed a whole range of other microbes when they die, working like plants to create complex and diverse ecosystems.

One of the coolest types of fuel for chemolithoautotrophy is pure radiation. When you hit water with radiation, it splits into hydrogen and oxygen, which are two of the best chemicals for feeding chemolithoautotrophs. Radiation is something we know about from bombs or from nuclear waste. But there are also natural forms of radiation around us all the time. The energy involved in the formation of the elements billions of years ago charged up radioactive elements like thorium and uranium, which decay spontaneously but very slowly. Consequently, all sediments everywhere constantly push out tiny amounts of radiation—too tiny to do any damage to us, but enough to split water and fuel microbial life. One of the deepest microbes ever found, discovered in a commercial mine 2.8 km deep,* is called

---

* The lead author on this study, Tullis C. Onstott, explained to me that it's hard to know the exact depth that these microbes are from. The scientists descended to 2.8 km in a gold mine to retrieve fluids from fractures in the rock, but the source of the fluids shooting out of the rocks could have come from much deeper.

Desulforudis audaxviator; its species name means "bold traveler." This organism is locked far away from any recent interactions with the surface world, possibly subsisting entirely on chemicals produced from radiation.[1] Seems like a good gig, living right next to your food trough. The only drawback is that the amount of radiation is so low that the food delivery rate is very slow. This lifestyle is only suitable for a committed sloth, so it's tailor-made for an intraterrestrial.

Some intraterrestrials combine subsurface chemicals with oxygen and other oxidized compounds from the surface world. But other intraterrestrials, such as methanogens and acetogens, can survive completely on chemicals from Earth's crust, so these guys don't need to bother with the sun at all.

To appreciate how incredible this is, consider what would happen if the sun were extinguished today. All the photosynthesizers, like houseplants and algae, would die within a few years. Other life on the surface would keep going for a while, eating leftover plant matter. Humans, I'm guessing, couldn't make it more than a few years past the plants. There might be some cockroaches who could last another hundred years or so, but that's probably it for the animals. Fungi and bacteria that depend on plant matter and oxygen would probably last a few million years longer, but after all the food and oxygen was gone, they too would die. The intraterrestrials, you may have guessed by now, would be just peachy. They probably wouldn't even notice that anything was amiss. In fact, even the intraterrestrials who normally rely on oxygen deliveries from the surface might avoid asphyxiation since some intraterrestrials *actually make oxygen in small amounts*.[2]* None of these little guys care whether

---

* This is a fascinating, newly discovered metabolism that may change our fundamental understanding about where oxygen gets made on this Earth.

the sun rises in the morning or not. They're buried too deeply in Earth's crust to notice.

If the organisms in the subsurface ecosystems I've discussed so far seem nearly autonomous from the surface—unconcerned with even the survival of the sun—the organisms in the second type of subsurface ecosystem are the opposite. They are entirely dependent on leftovers from our surface world, like societies living on islands supported by food brought from the mainland. The largest subsurface environment that is dependent on food falling from above is marine sediments. Marine sediments can be tens of kilometers thick, and they underlie all the world's oceans; a serious amount of space for life to occupy. These sediments build up from the constant rain of gunk from the oceans or runoff from land. This ooze* serves as nutrients, energy sources, and solid substrates for subsurface microbes. At the end of a big river like the Ganges or Nile, these sediments pile up quickly. Here, a particle might only spend a few hundred years sitting on the seafloor before it's buried underneath more material raining down on it.

In other parts of the ocean, like in the middle of the Pacific and Atlantic Oceans, however, the seafloor only gets a light dusting. No continents are nearby with rivers spraying like muddy firehoses. Even the occasional dust storm from the Sahara or Gobi deserts doesn't reach them. Not only does being far away from land diminish the pileup of sediments, it also reduces the amount of plankton on the ocean's surface—plankton that could've rained down after the long process of being eaten, pooped, and recycled as they fell through the ocean's depths. This is because, in addition to sunlight, photosynthetic plankton

---

* This is not onomatopoeia. It's a technical word. It means the leftovers from the dead bodies of organisms that settle down through the oceans onto the seafloor.

need iron, which mostly comes from rivers and wind-driven desert dust. The middle of the ocean is thus a gorgeous waste-land with hundreds of kilometers of sparkling clear blue water. In these areas, a particle might sit at the seafloor for thousands of years or longer before enough sediment piles up and pushes it into the subsurface.

Even though the sediments themselves were delivered by riv-ers and falling plankton, many of the intraterrestrials that live there, improbably, were not. The microbes we find buried deep in marine sediments bear little resemblance to the ones found in rivers and oceans. They are their own thing, even if they're eating packed lunches from somewhere else.

You may be wondering at this point what, exactly, samples of these exotic environments look like when scientists pull them from the ocean floor. Visually, they're often underwhelming. They look like mud, rock, or fluids, which can feel like a bit of a letdown, given how difficult they are to retrieve. I am occasion-ally contacted by producers or writers of scientific TV shows and documentaries who find the deep subsurface evocative (rightfully so). Our conversations inevitably get to this ques-tion: "Now, how would your samples look on film?" When I reply, "Oh, the samples are mud. They look like . . . mud," the person usually tells me they've enjoyed talking to me and they'll get back to me later.

But the fact that it's commonplace is precisely why Earth's subsurface biosphere is so compelling. Mud is everywhere, which means it is important. If you add up the total amount of mud* underneath all the worlds' oceans, you come up with a

---

* Actually, I should say sediments because "mud" is a technical term that implies a certain grain size. But I use "mud" here to emphasize that it feels like sludgy stuff that we're all familiar with.

volume equivalent to about the entire Atlantic Ocean.[3] And, per cubic meter, there are 100 to 100,000 times more microbial cells in mud than there are in seawater.[4] That means that there's so much intraterrestrial life in the subsurface that it's hard to even fathom it. The total amount of microbial cells in the marine sediment subsurface is estimated to be 2.9×10^{29} cells.[5] This is about 10,000 times more than the estimated number of stars in the universe. But that's not the *whole* subsurface. You'd have to at least double this number to include the microbial cells living deep underneath the land.[6] And some of these cells may have found pockets where the food is more abundant than the average location, so more cells can live there than our models predict. For these reasons, the actual number of microbial cells in the subsurface biosphere is certain to be much higher than our current estimates.

Importantly, the two types of subsurface environments I discussed in this section are not mutually exclusive. For instance, recall the plastic cores of mud that my colleagues and I collected using the robotic arms of the *Johnson-Sea-Link II*, as we sat at the seafloor methane seep. When I examined those samples back in our laboratory at the University of North Carolina, I discovered that buried in this goopy black mud were bacteria and archaea that eat food raining down along with the sediments from the upper ocean. No surprise there. Curiously, though, I also found microbes that eat methane and hydrogen bubbling up from ancient petroleum deposits.[7] These microbes sat right next to the ones eating food falling from above. The amazing thing is that this crossroads of surface and subsurface food sources supercharged the microbial population, making the communities robust and diverse. Is it possible that many more such hot spots are waiting to be found throughout Earth's crust?

Perhaps, but in sediments that don't have a good source of chemicals, the environment changes drastically as you go deeper. Once you get a few meters deep, the situation is universally bleak: buried sediments no longer receive fresh food deliveries. Instead, the microbes must find creative ways to continue eating the small amount of food with which they were buried. It's akin to being given a lovely picnic lunch and being told to make that food last for the rest of your life. Shockingly, subsurface life is somehow able to make it work. How do they do this? We aren't entirely sure. That's partly because, as we'll see in the next chapter, cracking into solid Earth to get ahold of these samples is way harder than it looks.

2

CRACKING INTO SOLID EARTH

ONE THING is true for all subsurface environments, whether the food comes from above or below: they're hard to sample. You usually can't simply walk up to a site and start shoveling. There are rarely stairs or elevators leading down into them. Even in seawater, tossing sampling gear off the back of a ship requires technology, skill, and luck. Accessing the subsurface always requires effort.

This chapter is about how scientists obtain subsurface samples—from places shallow enough to reach with our hands, deep enough to require launching a multinational drilling operation, or fortuitous enough to catch microbes as they rush up to the surface on their own in natural springs. No matter how we do it, though, sampling the subsurface always involves at least a little bit of adventure and, as I'll describe, sometimes a bit of pain and agony. We don't do this job because it's comfortable.

Sampling Down to Thirty Meters

One of the least difficult ways to reach down into Earth's crust is to poke a handheld plastic core tube into mud and pull it back out with sediments stuck inside. It's like poking a straw into a slushy drink, sticking your thumb on the top, and pulling out the slush. This straw-poking method (which works equally well when I deploy plastic cores by hand in a North Carolina estuary as it does when I deploy them by submersible robot arm in the ocean) doesn't penetrate very deep, but sometimes that's good enough for our research. As long as the sample comes from a place that's reliably dark, with very little influx of fresh nutrients from the surface world, it is in the subsurface. One place that *always* fits this description is the bottom of the deep ocean. Hundreds of meters underwater, whatever you pull up from the muck is going to have been permanently removed from sunlight, with only a vestigial surface influence.

Slight modifications to a hollow core allow us to use the straw-in-a-slushy method to go even deeper. This modified method, called gravity coring, involves converting a hollow pipe into a giant, thirty-foot-long arrow by attaching metal stabilizing fins and a couple tons of metal weight to its head. The gravity core is cranked out over the water by a research ship's crane and then released. It shoots like an arrow straight through the water and thwacks into the seafloor, ramming the pipe deep into it. The filled pipe is then heaved out of the seafloor by the pulling power of the ship's winch and cranked back up onto the ship, taking care to make sure it doesn't swing around and kill anyone when it comes onto the deck. The inner plastic pipe, or core liner, slides out of the metal pipe and is laid out on the deck where it can be split lengthwise with a circular saw. You can then look across these layers and sample wherever you want.

When I look at cross sections of these giant cores, I am always struck by the fact that, even though they're "just mud and rocks," the different layers of seafloor have subtle variations that can be quite beautiful. The sediments shift between tan, white, red, gray, black, or even bright blue or seafoam green, sometimes in layers or tiger stripes. There can be small crusty bits of iron rust or delicate carbonate crystals that show that microbes have been hard at work making minerals. Other times, there are gravelly or shelly layers, hinting at what the seafloor was like many thousands of years ago. In 2009, I leaned over my pregnant belly to pull fluids from a gravity core as I swayed on the deck of a ship in Aarhus Bay, Denmark. As usual, as I sampled deeper into the core, pulling water out became more difficult because the sediments had compacted over time. But then I hit an unexpected gravel layer that was full of water. I did a quick calculation and estimated that these coarse-grained sediments were probably laid down when the Vikings ruled Denmark about a thousand years ago.

The sensory experience of examining a core isn't limited to sight. It can be even more informative to smell it. Right at the top, it may not smell like much at all—maybe like clay you would use to make pottery. But as you move your nose over the deeper parts of the core, it will start to smell like rotten eggs. The smell comes from sulfide, which is a waste product from microbes that breathe sulfate. Cores from the White Oak River estuary I often visit in North Carolina don't start to get smelly until about four inches down. In the deep sea, it can take meters before things start to get stinky. The difference is because an estuary gets a lot more organic matter than the deep sea does, so processes go faster there. Often, the sulfide smell is mixed with other microbial byproducts from breaking down dead stuff. These smells are molecular cousins of the ones that give subtle scents to fine wines and cheeses, though I promise you that marine sediment smells considerably less pleasant. Nevertheless, my wonderful PhD advisor, Andreas Teske, would waft the vapors toward his nose, like a mud sommelier, discussing the delicate top notes of cadaverine, with a rich body of putrescine.

As you travel farther down the core and further back in time, the sulfury smells slowly subside. Sulfide is highly reactive, and one thing it likes to react with is iron, which is everywhere in the subseafloor. Most of the microbes floating around in the oceans are scrimping, saving, and fighting with each other to suck up every last atom of iron. But down in the sediments, iron accounts for a whopping 8 percent of the total weight of material, making intraterrestrials positively wealthy with iron. This iron quickly binds up all the stinky sulfide and makes black pyrite. Even further down the core, the black pyrite deposits disappear, leaving clays and carbonates that often smell chalky, and sometimes smell like a gas station. In places like the Gulf of Mexico, the same ancient hydrocarbon petroleum deposits that attract companies like Shell and BP make our samples smell to high heaven.

Sometimes coring is dangerous. Methane is not very good at dissolving in water. Like the bubbles in a Coke bottle, methane wants to fizz out. However, when it's placed under an ocean's worth of pressure, methane will dissolve. As a result, mud at the bottom of the ocean is often overloaded with methane. When we take it out of the seafloor and drag it onto the back of a ship, the sharp pressure drop can make methane explode like a bomb. In 2006, I was on a ship in the Gulf of Mexico. We pulled up core after core, hoping for methane, but not finding much. We dutifully worked around the clock analyzing the cores and processing samples, even though they weren't quite what we wanted. Finally, we did yet another cast of the gravity core and when we hauled it up on deck like an elongated metal fish, mud shot out of the top of the pipe like a giant fountain. We wanted a manageable amount of methane, but instead we hit the mother lode and it exploded. I watched my beautiful samples spray across the back deck of the ship and thought, "Well, I got what I wished for."

Sampling Down to Kilometers

What if you want to drill extremely deep into the subseafloor? Like hundreds of meters into it? For that, you can't just toss a heavy pipe off the back of a ship, no matter how many tons of steel plates you attach to its head. Instead, you need some industrial-strength drilling equipment, for which you need a lot of money, for which you need an organized national or international effort.

Currently, scientific deep subseafloor drilling is organized around three main national groups from the US, Europe, and Japan; and deep continental drilling is led by one big international effort. These programs support three vessels: the RV *Joides Resolution* (US), the RV *Chikyu* (Japan), and ships retrofitted with drilling equipment to fit into tight spaces (Europe). The RV *Joides Resolution*, which has historically been the workhorse of scientific ocean drilling, can drill boreholes about a thousand meters into the seafloor, depending on the water depth and the type of material that makes up the seafloor. Its high-powered stabilizers keep it stock-still at the ocean surface, even as currents and waves try to toss it about. This stability is necessary so that hundreds or thousands of meters of "drill string," which is actually a hollow metal pipe, can feed out below it, sticking the ship to the seafloor like a giant floating lollipop. Then a massive drill bit is fed through this pipe to drill into the mud and rock, using seawater for lubrication.

Unfortunately, this seawater carries its own microbes that can contaminate the samples. To avoid this issue, drilling teams try to use Advanced Piston Coring (APC) whenever the mud allows it. For APC, a high-powered stroke is detonated at the bottom of the borehole, jamming nine meters of pipe into the seafloor in less than three seconds, so it doesn't contact drilling fluids. As further precautions, we only work with the center of each core and add a volatile substance to the drilling fluid so we can periodically test the core material to make sure no drilling fluid seeped in. This caution is well warranted: if you're drilling a hole hundreds of meters deep into sediments under miles of seawater, many tens of these pipes come up to the deck full of precious subsurface sediments that haven't seen the ocean, much less the light of day, for hundreds of thousands or millions of years. We use these volatile tracers to know if we accidentally got a bit sloppy with the drilling and contaminated everything with normal seawater microbes.

Sadly, the RV *Joides Resolution* will soon be decommissioned, which means that it will no longer be used for scientific expeditions. At the time of this writing, the US has no concrete plans to replace it, and so will soon have no way to drill into the seafloor for public scientific exploration. Private companies will continue to do so, plumbing the economic riches of the subseafloor, but society will benefit from these missions only to the extent that companies are willing to sell us what they find. I am, however, hopeful that our societies will see the value in scientific drilling and continue to support it.

The Japanese ship *Chikyu* was designed to do something that had never been done before: drill all the way through the crust to the mantle. Remember when I said no one has ever drilled to the mantle before? Well, the *Chikyu* is the vessel poised to do it. Though it has not yet achieved this previously unthinkable feat, to date *Chikyu* has enabled scientific exploration of life up to 1.2 km into the seafloor off the coast of Japan in Expedition Leg 370. It will have to go a few km deeper when it eventually manages to drill through the thinnest part of the crust to get to the mantle. The scientists of the Leg 370 mission, who come from Japan, Germany, Denmark, the US, the UK, Canada,

Switzerland, Australia, and Korea, showed that life thrives this far down into marine sediments. Their samples revealed blooms of microbes in sediments that had been buried for many millions of years.[1] Even in this very remote section of the deep biosphere, it appears that the subsurface still has something to offer in the form of warmth, which cooks the buried food and makes it more nutritious. As only a few expeditions have ever drilled deep enough to observe this phenomenon of reactivation of food sources, we don't yet know how common it is on Earth.

Despite these impressive efforts, the deepest life on Earth has not been found by drilling into the seafloor. The deepest life on Earth has been found by scientists boarding terrifyingly rickety, sweaty-hot elevators and plunging deep into the Earth in commercial mines. Heavy metal industries, such as goldmining, dig tunnels many kilometers into Earth's crust. Since the deep Earth is hot, mines are hot, warmed by the slow radiation of Earth's primordial heat of formation. Scientists who forge good relationships with the mine operators can sometimes travel down into these mines to siphon fluids from fractures deep in these rocks. As with marine sediments, no matter where scientists look in these deep rocks, they find abundant life. One of the biggest finds from such work is the bacterium called *Desulforudis audaxviator*, which I mentioned in the previous chapter, that eats the reaction products of pure radiation and water. Yum. And, as you may recall, it was found 2.8 km deep in a South African gold mine.[2]

Catching Subsurface Life as It Bubbles up to Us

If you find yourself without any submersibles, gravity cores, drilling vessels, or the good will of a commercial mining operation, fear not. There is another option for sampling the subsurface: you can wait for it to come up to you. In the previous chapter, I described the many parts of Earth where tectonic plate movements, salt deposit buoyancy, or heat convection of subsurface waters sloshes around the chemicals that fuel chemolithoautotrophic life. As luck would have it, some of these microbe-laden deep waters shoot all the way to the surface. Right where we can catch them.

In the ocean, these jets of subsurface water are called hydrothermal vents, and they often form chimneys of mineral precipitates, as the material that is dissolved within the fluids reacts with cold oxygenated seawater.* These hydrothermal vents are like natural sampling ports for deep subsurface fluids.[3] Of course, to sample vents at the bottom of the ocean, you need to access them with a submersible. Then, the next hard part is finding a way to sample these vents using the submersible's robotic arms without sucking in a bunch of seawater along with it. After having worked with oceanic samples like this for years, my colleagues and I recently decided that we'd see if we could try similar sampling techniques on land.

On land, when water shoots to the surface, we call it a natural spring. These are often hot, because the interior of the Earth is hot. The accompanying carbonate or silicate precipitations often form large mounds or terraced fields around the vent, like the ones made famous in Yellowstone National Park. One advantage of sampling on land is that we don't have any pesky seawater to keep out of our samples. We just push the pure fluids directly through filters to catch the microbes. Also, the logistics of working on land are a million times easier than working underwater. I cannot overemphasize how much easier it is to use GPS to drive up to an easily visible hot spring and sample it with your own

---

*You can see them in high resolution in James Cameron's documentaries *Aliens of the Deep* and *Volcanoes of the Deep Sea*.

gloved hands, than it is to find a hydrothermal vent at the dark bottom of the ocean and maneuver a slurping tool with clunky robot arms. The disadvantages of sampling on land have more to do with humans. More than once, I've walked up to a hot spring and discovered a family sitting in it. I awkwardly ask them to scoot over, so I can try to get close to the water source. Unsurprisingly, these samples often contain a mixture of DNA from the subsurface and human contaminants.

The second disadvantage to working on land is less about the science itself and more about the scientists. Without the protection of a strong titanium sphere, we frequently come face-to-face with harsh conditions. High altitudes can be life threatening, and driving in the back country of places like the altiplano in Argentina is not for the faint of heart—we found our teeth chattering against our brains as we slammed into deep ruts and small boulders. I discovered this firsthand the first time I visited Argentina in 2019.

The area we were heading to was above 4,000 meters (13,000 feet), and many of us were morbidly curious to see what the altitude would do to our bodies. The lower oxygen partial pressure means just getting out of a chair can make you gasp for breath. Doing anything mildly athletic is impossible. But the real problems are the unpredictable side effects. Your hands might balloon into sausages. Your head may feel like you've been on an all-night bender. You may throw up once, twice, or uncontrollably. You may turn pale and be unable to lift your arms or stay upright. You may get stupid. Once, while visiting the high-altitude Mauna Kea observatory in Hawaii, I stood around stamping my feet, freezing and cursing myself for not packing a warm jacket. When I got back down to sea level, I found my jacket right at the top of the backpack I had been wearing. I had simply forgotten it was there. This foolishness occurred after being at Mauna Kea for merely a few hours. In Argentina, we were spending two weeks at altitude.

We piled into a small hotel in San Antonio de los Cobres at 3,700 m, ate some dinner, and collapsed into simple shared rooms. The next morning, as I entered the shared dining area, I discovered that every person in the group had endured the worst night of their lives, tossing, turning, and waking up every few minutes fighting to breathe. The breakfast room was silent and grim. I started passing out ibuprofen like it was candy, but it barely cut the sharpness of people's pain.

Our agony was partially lessened by the promise of what lay ahead: we were going to take some of the first microbiological samples ever in this remote place. We raced our trucks across the roadless altiplano, using only GPS points and volcanoes for navigation. Compared to the jagged roads we had navigated to get there, the altiplano felt like flying. The Andes volcanoes, the tallest ones in the world, pop up in massive, civilization-ending eruptions, and then slowly erode over hundreds of thousands to millions of years. The resulting eroded rubble settles in the basins between the volcanoes, leaving flat gravel behind. You couldn't pay a paving company to do a nicer job.

It's not just gravelly rocks that tumble down from the volcanoes; the occasional rain leaches out salts and deposits them in central basins. Over time, this briny rainwater evaporates, making many kilometers of brilliant white salt flats, called salars. As we drove past one of them, light pink dots appeared on the horizon. As we drew closer, I saw that the dots had wings and were standing on one foot. They were flamingos. Anyone who knows anything about South America will not be surprised by the presence of these birds. Flamingos are key to the cultural and natural history of the region. But I knew very little about South America before I made my first trip to Argentina, and seeing a flamingo, chilling on one leg without any green plants around, in the middle of a high-altitude salty desert, was a humbling reminder that there's a lot about the world that I don't know.

Spotting a hot spring in a desolate rocky environment is easy. Any hint of bright green on a hillside tells you where to go; there's so little rainfall that these springs are the only way to support anything with leaves. We spotted our green mini-oasis off in the distance and began slowly making our way up the hill. As few people traveled this rustic road, it was poorly maintained, and we had to stop the car every now and then so we could hop out and move small boulders out of our way. When we reached the spring, there was a little mud-and-thatch house near it, with llamas wandering around. We had not seen a human-made structure for hours, so it was disorienting to find a house. Even more surprising was the man who came bounding out of the home with a big smile on his face, excited that there were women in our group. He told us that he had not left his house in years and that only a few people had visited him during that time. He was pleased to share his hot spring with us and accepted our offer of a small monetary payment, a baseball hat, and a chance to talk, which he did with glee.

Once we had finished filtering fluid and collecting a few small tubes of sediments and gasses, we packed up quickly to get back to the village before nightfall. But first, we had to deal with the most essential of physiological necessities: finding a place to go to the bathroom in a flat landscape where the vegetation is ankle-high at best. I generally consider toileting out in nature to be a job perk—a mountain vista beats the cold tiles of a bathroom floor any day. But that day, as I moved away from my group to find a place to go, I couldn't shake a pack of about seven curious llamas. They followed me around until I finally announced to them, "OK, we're doing this," and then did my business while they stood in a semicircle chewing their cud and staring at me dumbly. On another of these trips, the only vertical relief we saw all day was a large black pipe carrying water to the remote villages. I thought, "Perfect!" So, I leaped out of the truck, squatted behind the pipe, and leaned my bare lower back against its warm side. As I hopped back into the truck, triumphantly rubbing sanitizer into my hands, something wasn't right. My lower back was burning like crazy. The pipe was made of fiberglass, and the little needles were cutting into my skin with every jolt from the truck. So much for my brilliant workaround.

Workarounds, it turns out, are the mainstay of fieldwork, especially when you're working in a country with an inflation rate above 45 percent. Argentina's financial volatility required us to make on-the-spot cash transactions. After three days at that little high-altitude hotel in San Antonio de los Cobres, it was time to move on to the next sampling point. To pay the bill, we covered the hotel's reception desk in thick stacks of cash, thanked the proprietors for the nice stay, and made our way outside as they bagged it all up. When we regrouped out on the street, we realized we'd given the hotel our last bills. Because withdrawing money from our US banks came with a hefty 25 percent surcharge, Agostina Chiodi, the geologist at the University of Salta who was leading the trip, offered to withdraw money from her personal bank account to cover our whole group. We stood around in the street cracking bad jokes and kicking dust while she went into the bank to drain her account. Almost immediately, she walked back out with a stricken look and said, "The bank is out of cash!" We were screwed. We were in a remote village, and we had no way to pay for food, water, and a place to sleep. So, you know what we did? We walked our contrite asses back to the hotel and begged for our money back. Miraculously, they gave it to us! Agostina had stayed with them previously for fieldwork, so they trusted her to pay them back, which she later did.

In the end, we collected samples from twelve hot springs scattered all around the Puna region of Argentina. We froze each one of them immediately in a liquid nitrogen dry shipper—a lawn-mower-sized metal container whose walls can be chilled to the temperature of liquid nitrogen—that we carry with us on our trucks.* To our knowledge, no one had ever examined the intraterrestrials of these sites before, much less placed them in their geological context. But here we were, a bunch of geologists who were trained to think that biology was just green crud you wipe off rocks, and a bunch of biologists who were trained to think that life requires sunlight, working together to sample a world where neither was true. But if you only read the scientific papers we produced from this work, you'd never know what it *really* took to get these samples. A heavy dose of a spirit of adventure is the only way to make the science happen.

The scientific work really ramps up when we get our hard-won samples back home to our laboratories. If these microbes were more charismatic, we'd probably spend a lot of time gazing at their cute little feelers and frustules under a microscope. I, for one, think most microbes are adorable. But not *these* microbes. These microbes are tough, tiny, starving, and stressed out. To study these weirdos, we have to exercise some creativity.

---

*I could write another book about what it takes to get liquid nitrogen dewars through international airports, but it would be very dull, with lots of waiting, smiling, and trying to look nonthreatening. The fact that our equipment is technically approved by the US Federal Aviation Agency means nothing to the airline check-in attendants in Buenos Aires, staring in horror at our giant, fuming, mad-scientist-looking contraption.

3

THE TWO DNA REVOLUTIONS

INTRATERRESTRIAL RESEARCH would be much easier if we could spread bits of our freshly collected hot-spring fluids or deeply drilled mud across petri dishes and watch intraterrestrials sprout like mushrooms. Then we could pick them up with sterile toothpicks and do experiments on them to see what makes them tick. Researchers have tried something like this many times, and inevitably they get some microbes to grow. But even when they keep the fluids or mud under high pressure, hoping to approximate the intraterrestrials' home environment, the microbes they grow are rarely the same ones we see in nature.

So, if we can't grow them, how do we even know we're missing them? The answer is that we found them through the tools of the DNA revolution. This revolution blew our whole field of science wide open. In this chapter, I will describe what the DNA revolutions (actually, there were two of them) have looked like from my front-row seat, and how they have allowed us to gradually uncover the shocking enormity of what we've been missing about life on Earth.

The First DNA Revolution

DNA stands for deoxyribonucleic acid, and it is the main informational molecule in all living cells. Almost every function a cell performs is encoded in its DNA, which is composed of only four bases (named A, G, T, and C) strung together like beads on a string or words on a page. These "words," or genes, are important because they direct the construction of molecules like RNA and proteins such as enzymes. These biomolecules can then do work inside each living cell.

DNA sequencing, the process by which we read the order of thousands of these bases, involves putting physical strands of DNA into a machine that reads them and converts this information into a computer file. When I was a graduate student in the mid-2000s, the field of biology was right in the middle of the first DNA sequencing revolution. Prior to this revolution, the only way to sequence DNA was to painstakingly analyze each base by hand, through laborious methods that were radioactive and other kinds of toxic. If we had not been able to progress past this slow and dangerous method, we'd never have discovered the wide diversity of life we now know exists.

The first DNA sequencing revolution was driven by two breakthroughs: capillary electrophoresis–style Sanger sequencing and the polymerase chain reaction (PCR). Sanger sequencing is an excellent method for determining long, high-quality DNA sequences. There is, however, a significant catch: it only works if you have a ridiculously large number of exact copies of the same DNA sequence, and *only* that single DNA sequence.

For many years, this was a major limitation, as nature is messy and complex. Any spot of soil the size of a pencil eraser contains billions of microbes. Each microbe has at least one genome inside it. Each genome has at least a few thousand genes. Each gene has hundreds to thousands of As, Gs, Cs, and Ts. Add it all together and there are thousands of billions of DNA strands pretty much everywhere, thanks to the ubiquity of the microbial world. How, then, do we extract one specific strand of DNA from this morass?

The first thing we do when we get a sample back home is physically isolate the DNA from whatever gloppy muck we pulled up along with it. All the fats, sugars, proteins, and bits of soil and mud have to be cleaned out. Then, to extract DNA, we break the microbial cells open so the DNA falls out of them, add chemicals to prevent the DNA from sticking to the muck around it, and then clean and purify the DNA. Though there are several different methods for performing each of these tasks, testing every variation of every step of the process is impossible, so when we find something that works, we just go with it and don't ask questions. One of my colleagues, Kostas Kormas, a scientist at the University of Thessaly, Greece, was once so exasperated at failing to extract DNA from his hydrothermal vent fluid samples that, in a fit of anger, he cooked them in a microwave. To his surprise, it worked! In other cases, a perfectly sensible process turns out to be a total disaster: I once spent a whole summer extracting DNA samples only to discover that every time I went through the last step of the process, I had been unwittingly washing most of my DNA down the drain. When I realized what I had done, I cried, and gathered friends at the bar for an impromptu wake for my three months of work.* In general, DNA extraction is one of

---

* I also began plotting how to publish my failure as a manuscript, which I later did. (Lloyd, K. G., MacGregor, B. J., & Teske, A. Quantitative PCR methods for RNA and DNA in marine sediments: Maximizing yield while overcoming inhibition. *FEMS Microbiol. Ecol.* 72, 143–51 (2010). Spite is one hell of a performance-enhancing drug.

the least precise things I do as a scientist, and if I think about it too deeply, I feel queasy.

When I was working on my PhD project in the early 2000s, whenever we made it through the woods of DNA extraction, the next step in the process involved a neat trick with *Escherichia coli*. More commonly known as *E. coli*, this bacterium is often transmitted to humans through unwashed salads and can result in bloody diarrhea, vomiting, and an intense desire to die, if not actual death. It grows like wildfire and has a habit of picking up genes from its environment that make it an agile adversary against our immune system. There are, however, strains of *E. coli* that can't infect us, yet they still grow fast and pick up DNA that's lying around. These good-witch *E. coli* will kindly accept a DNA sequence and make billions of copies of it, which we can use for Sanger sequencing. This method is called making "clone libraries" because each DNA sequence comes from a single clone of *E. coli*.

Of course, if we gave the *E. coli* all the DNA extracted from a crushed-up rock or a squirt of subsurface fluids, we'd end up with random bits of DNA that we may or may not care about. Instead, we have to be selective in the DNA we offer to these *E. coli*. To pull out just one gene from this genetic stew and make a bunch of copies of it, we first perform PCR. The invention of PCR was essential to the DNA revolution because it is the only way to get enough molecules that you can actually work with so you can learn about them. Unfortunately, the process involves a lot of fiddly enzymes and small molecules. If DNA extraction was anxiety inducing, PCR can turn people religious. I've seen more than one PCR machine surrounded by prayer flags, little altars, or lucky cats. Why all the fuss? Well, for PCR to work properly, you have to grab just the right gene, using a little DNA handle called a primer that consists of about twenty bases of DNA. You also need exactly the right mix of components and reagents. You know what ruins PCR? Too much salt, metals, and organic molecules. You know what also ruins PCR? Too little salt, metals, and organic molecules. Getting the balance exactly right is impossible for the simple reason that it is impossible to extract the DNA in a perfectly clean manner from all the crud that's around it. Even DNA *itself* can mess up PCR. So every PCR attempt is at least a little bit doomed, and one just hopes (perhaps even prays) that there is enough DNA in the sample to outcompete all these inhibitors.*

Despite the difficulties and occasional nightmares inherent to this overall process of PCR-amplifying a gene, cloning it into *E. coli*, and Sanger sequencing it, the method worked well enough to give the first inkling that the diversity of the microbial world dwarfs that of visible life on this planet.[1] There were, however, two main problems that limited the usefulness of this approach. First, the method was slow. As a result, we were forced to conduct a census of billions of microbial cells by looking at just a few tens of DNA sequences. It was like trying to do rainforest ecology if you could only see four species. Try guessing the total number of species in a rainforest from a list that only includes two types of trees, one monkey, and one frog. It's a statistician's nightmare.

This is roughly the state that the field of microbial ecology found itself in during the early 2000s. We identified such a flood of diversity from nearly every environment, using PCR primers and clone libraries, that we were lulled into a false confidence that we had caught them all. But we had not caught all the microbes out there[2]—not by a long shot. For that, we needed a second DNA sequencing revolution.

---

* My colleague Emily Fleming-Nuester at Chico State University jokingly claims that PCR stands for Pure Chance Reaction.

The Second DNA Revolution

While I was laboriously making clone libraries, the pressure to sequence more DNA, faster, with lower error rates, at ever-decreasing prices per base was driving a multibillion-dollar industry.* The first big leap was made by a company called Roche, which used a method called 454 pyrosequencing to produce tens of thousands of individual DNA reads per sample. The new method cracked the hunt for new types of microbial life wide open, as scientific progress was no longer limited to the number of *E. coli* globs a beleaguered grad student could stab with a toothpick.

In the mid-2000s, Julie Huber and Mitch Sogin at the Marine Biological Laboratory used this new 454 technology to sequence genes from ~10,000 microbes in a liter of deep-sea water.[3] To give you a sense of how far ahead of the rest of us they were at the time, that same year I published a paper with genes from only ~50 microbes per sample.[4] Julie and Mitch discovered that seawater contains thousands of microbes that are undetectable with the Sanger technology because they are in very low abundance. Each one of the members of this so-called rare biosphere was poised to grow into a more abundant population under certain conditions, with each responding to a different environmental shift. This caused Julie and Mitch to make a scandalous conclusion: there were millions more types of microbes out in nature than we'd ever guessed there could be. When they first presented these results at a conference, I overheard one

---

* Microbial ecology doesn't tend to have the financial backing that drives big technological advances like this on its own. But human genome research does. Thanks, guys!

attendee loudly telling everyone in earshot that it was all "Bullshit!"

Initially, the reason why rare biospheres were controversial is that 454 pyrosequencing was error prone. This is also the reason I'm not completely embarrassed that I was still using the old *E. coli* Sanger method in 2006. This 454 technology depends on immobilizing single-stranded DNA on a surface. DNA does not like to be single-stranded. It likes to be a double-stranded helix. Whenever two strands of DNA are pulled apart, it's like a bunch of bases yelling "Marco . . ." and waiting for someone to shout back ". . . Polo" so they can find each other again.* Except in this case, the single-stranded DNA is shouting, "AGTCCC . . ." and waiting for ". . . GGGACT" to come stick to them. This is because As and Ts want to go together, and Gs and Cs want to go together. To get the order correct, I refer you to Missy Elliot's famous line, "You've got to flip it and reverse it."†

In 454 pyrosequencing, every time a "Polo" binds a "Marco," a flash of light is emitted, which is used to determine the sequence of the DNA. Making copies of DNA does not emit light on its own; if it did, our bodies would sparkle all the time, and, sadly, our universe is not that rad. The flash of light comes from a cocktail of enzymes that are part of the pyrosequencing process, one of which comes from the butt of a firefly. The cool thing about this technology is that it automates the process: millions of different DNA strands are immobilized across a tiny

---

* There is a children's game common in the US, where one child in a swimming pool closes their eyes and tries to catch the other children in the swimming pool by yelling "Marco" and having the others yell "Polo" in response. It also works with inebriated adults.

† From Missy Elliott's hit song "Work It," 2002. For DNA, this is called a reverse-complement, but if DNA had been discovered after 2002, I assume we would call it a "flip-reverse" instead, because that song is great.

chip, a bunch of one type of base (for example, G) is washed over all of them, a high-resolution photodetector records which DNA strand (identified by where it is on the chip) had a flash (meaning it took up a G), then the same thing happens with the next base, and so on, cycling through the four bases until it's done. The uncool thing about this technology is that it fails when it encounters a DNA sequence with a bunch of the same base in a row. Given this limitation, many scientists found it difficult to trust that all of the rare sequences that Julie and Mitch were finding were real. They could have just been errant flashes of firefly butt enzymes.

This skepticism was put to rest by a company called Illumina. As with 454, their process also immobilizes DNA and amplifies it in one place, such that a small population of DNA strands will emit flashes of light. But in Illumina's process, the flashes of light are more carefully controlled—only happening one at a time, even if you have a bunch of repeated bases in a row. This solved the major worry about 454 pyrosequencing and gave definitive confirmation of the existence of the rare biosphere. Nevertheless, Illumina's process is not 100 percent accurate for the simple reason that it involves mapping the locations of millions of flashing lights on a chip the size of a postage stamp, which is hard to perfectly do every time. However, any small lapses in quality can be overcome by the fact that this process makes such a massive flood of data that scientists have the luxury of throwing away the imperfect DNA sequences, keeping only the most pristine.

What the Second DNA Revolution Revealed

Thanks to the second DNA revolution, scientists no longer need to estimate all the species in a metaphorical rainforest based on a few metaphorical monkeys and trees. Now, we have millions of individuals to help guide our census. Nevertheless, these new powerful datasets have resulted in wildly divergent estimates of the total number of species of microbes on Earth, ranging from a million to a few trillion species.[5] These various estimations have made two things crystal clear. First, regardless of which estimate one accepts, there are astronomical amounts of microbial species out in the world, many more than we knew existed twenty years ago; second, all these estimates are *wrong*. Every single one of them.

That's right, I said it. They're *all* wrong. The reason for my confidence is that all of these estimates have a glaring central flaw: PCR amplification. Remember PCR? That whimsical process that brings strident atheists to their knees?

The biggest problem with using any of these DNA sequencing methods to make a global microbial census is that these methods require us to use DNA primers to grab the DNA sequences we want. To assess the total census of all the microbes present in a given sample, we have to make a guess about the correct sequence for a DNA primer that matches every species on Earth. In essence, primers impose a pre-selection based on the assumption that we already know the DNA sequence we want to find. That's not a great recipe for discovering something new.

To go back to the rainforest analogy, using primers to capture all the diversity would be like trying to describe all the life in a rainforest while only being able to detect things with wings. There are a lot of different winged things in a rainforest—birds, insects, bats, flying squirrels—even some plants have wings on their seeds. In fact, there are enough winged things that a researcher might be fooled into thinking that this method gave them a fairly good understanding of the total diversity in the rainforest. In reality, however, the researcher would remain ignorant about all the wingless beetles, ants, plants, fungi, mammals, and reptiles that are in the rainforest. They would barely scratch the surface of the total diversity, and, worse yet, they wouldn't even know it. They would be blind to their own ignorance.

Since all the current estimations of total microbial diversity on Earth are made with methods involving DNA primers, I can say with great confidence that we have not, in fact, reached the correct answer. Luckily, the new DNA sequencing technologies have another advantage. They produce so many DNA sequences that we can skip PCR altogether and sequence every bit of DNA we extract, thereby offering a path to a more accurate census of life on Earth. The reason a new non-PCR census hasn't been performed yet is that the new non-PCR way to sequence DNA produces taxonomic information in smaller quantities than the PCR method. Since we're not targeting a single taxonomic gene, we only get a relatively small number of taxonomic marker genes in a larger pool of all the other genes in a sample. But eventually the new methods should catch up.

Even more importantly, this new high-throughput DNA sequencing brings with it another benefit, perhaps the most powerful outcome of the DNA revolution: the ability to predict the lifestyles of these strange types of microbes. DNA can tell us whether or not a microbe is capable of building its body out of pure carbon dioxide, like plants do, or whether it has to eat other organisms, like animals do. It can tell us how a cell moves, builds external membranes, defends itself, brings nutrients into the cell, spits out wastes, breathes, responds to stress, makes antibiotics, and much more. We can use these DNA sequences to predict what food intraterrestrials eat, or whether they are likely to form spores—hardy structures that keep cells just barely alive in inhospitable conditions. Of course, we can't know for sure that these genes are doing exactly what we think they're doing. Even though an intraterrestrial's gene may look like a well-known gene, it might work differently for the intraterrestrial. There are, however, ways to confirm our predictions. We can, for instance, turn a gene into a protein and check its function in a laboratory. Or, better yet, we can look for direct evidence of these functions in nature. For instance, if the genes indicate that an intraterrestrial makes methane, we can go back to the environment and see if there's methane. In this way, we can piece together the likely lifestyles of these intraterrestrials from DNA and environmental data alone.

Nowadays, when we return from a hot spring or a drill ship with a slew of frozen samples, we just extract the DNA and send it to a sequencing center that sends us back millions of short bits of DNA, randomly scattered across all the genomes of all the microbes in our sample. For my graduate students, the tedious part of the work does not involve picking *E. coli* colonies, like it did for me—it's bioinformatics. Bioinformatics is a catchall term for analysis and interpretation of terabytes worth of tiny snippets of DNA sequences, which mainly involves using computer algorithms to piece these tiny sequences back into the genomes they originally came from.

We're forced to sequence the DNA in small pieces because that's what the current Illumina technology requires. But wouldn't it be nice if we didn't have to work to put DNA sequences back together after purposely ripping them apart—if we were able to sequence the DNA while it's intact? In addition to this short-sequence problem, another drawback to current DNA sequencing technologies is that they don't really look directly at the DNA we give them. Instead, they only "see" the DNA while making copies of copies of copies of the original piece of DNA. If DNA sequences were books, our current DNA sequencing technologies would be like reading carbon copies rather than the originals. We use a natural enzyme called DNA polymerase to copy DNA, but DNA polymerase is, by necessity, a little bit incompetent at its job. It needs to allow the occasional wrong nucleotide to sneak in; otherwise we would never have experienced the mutations we needed to evolve. These enzymes' unreliable nature is great for life's ability to sprout feathers or a tail, but not so great for accurately reporting DNA sequences.

Luckily, there's a single solution to the tandem problems of being forced to work from cut-up DNA sequences and being forced to use copies made by an enzyme that has perfected the art of making slightly crappy copies. We just need to directly sequence very long stretches of DNA. That's a simple thing to write, but a difficult thing to do. The current solution is nanopore sequencing. Nanopore technology uses enzymes, not to make copies of the DNA as with the other technologies, but to unwind and separate the strands of DNA, which is fed through little holes in an electrically charged membrane. As each single strand of DNA is fed through one of the tiny holes, a machine can detect and record which nucleotides are going through, based on how they change the charge of the membrane when they pass through it—the first truly direct sequencer for DNA. The other nice thing about this method of DNA sequencing is that it can handle long strands of DNA. At least in theory, this technology could some-day sequence a whole genome in one long read.

This is not to say that the third DNA revolution has already arrived. This technology is still relatively new, and it has a worse error rate per base than the other sequencing technologies. However, the process is improving quickly, and I have high hopes that a day will soon arrive when I'll no longer need computer-intensive algorithms to put together what I've torn asunder for the sake of DNA sequencing.

The two stages of the DNA revolution I've experienced during my career have been game changers. They have freed us from the limitations of traditional laboratory-bound microbiology. As we'll explore in the following section of the book, what we've found with these new technologies has forever changed our conception of what life is like on Earth.

PART II

HOW DO INTRATERRESTRIALS CHANGE OUR BASIC NOTIONS OF WHAT LIFE IS LIKE ON EARTH?

4

HUMANS AND OTHER PLANTS

THE DNA revolution showed us that the number of species on this planet is sky high. But the DNA revolution can also tell us how evolutionarily distinct each one of these new beings is from what we knew before. Are we discovering the equivalent of a new type of monkey? Or are we discovering the equivalent of all vertebrates, many times over? In this chapter, I will describe the strange new types of life we've discovered in Earth's subsurface and even teach you how to find them yourself, if you're up for the challenge.

A New Branch on the Tree of Life

When I started my PhD research at the University of North Carolina in 2001, one of the first things I learned was that DNA was rapidly becoming a powerful tool in the hunt for subsurface life, opening lines of research that had been impossible just a few years prior. These were, however, still early days in the DNA revolution, and we had to pick which gene we were going to use to identify novel organisms.

How did we know which gene to pick? To answer this question requires a quick look into the nitty gritty of evolution. Every living cell has to copy its DNA to grow and divide, but the cellular machinery often makes mistakes* during the process. Evolution happens when these mistakes are advantageous, so that the offspring are more successful, passing the mistake along to their progeny. This stabilizes the newly mutated gene. But some genes don't tolerate this imperfect copying. For the genes that are crucial to life, any little oopsie is lethal to the offspring, and thus the mutation is not carried forward by the species. It is these types of genes—the stalwarts that resist evolutionary change—that we needed to investigate to determine where an organism falls on the tree of life. This is because their DNA sequences more faithfully record the evolutionary history of the whole organism, rather than the shifting sands of momentarily advantageous mutations.

For my advisor, Andreas Teske, and me, the choice was easy: 16S rRNA genes. These genes were originally used for microbial taxonomy by Carl Woese and his lab members in the 1980s and had come into widespread use by the early 2000s. What makes these genes so special? They encode the ribosome, which is essential to everything a cell does, so even one little change in the nucleotides (say, from an A to a T) in the wrong part of the gene might kill a daughter cell. As a result, the sequence stays more stable than the rest of the genome, which evolves around it. And while 16S rRNA genes are not 100 percent stable, their rate of change is slow enough that even after ~3.8 billion years of life on Earth, every living thing on the planet still contains some version of the gene. No creature has been able to get rid of it entirely, as far as we know.

---

* Some mutations are caused by mistakes when DNA polymerase makes imperfect copies, as I discussed in the previous chapter, but mutations can also occur for several other reasons: random bits of DNA will cross over with each other, enzymes will nick the DNA, or UV radiation of reactive chemicals will damage it.

It would make a thrilling scene if I told you I was staring through the eerie porthole of a submersible when I used these genes to discover a totally new form of life. But that's not how DNA sequencing works. It takes months of work after the cruise, when we're back up on land, to make it happen. In reality, I was sitting on a beat-up old chair in a windowless shared graduate student office when I downloaded files of 16S rRNA genes. And they looked something like this:

>DQ521765.1 Uncultured archaeon clone SURF-GC205-Arc3 16S ribosomal RNA gene, partial sequence

```
CCCGACTGCTATTTGGGTGAGGATAAGCCATGC
GAGTTGAATGGGAAACCTAAAGTTCCCATG
GCAAACTGCTCAGTAACACATGATCAACT
TACCCTATAGAGAAAATAACCTCGGGAAACTGAG
GATAATGTTTCATAGTTGAATTGGCTTG
GAAAAGTTTTTCGACGAAAGGGGTAAAAAAAATG
GTTTTTATCCGCTATAGGATAGGATCGTGTTCGAT
TATGGTTGTTGGTGAGATAATGGCTCACCAAGCC
GATAATCGATAGGGGCCGTGAGAGCGGGAGCCCCGA
GATGGGTACTGAGACAACGACCCAGGCCTTACGAG
GCGCAGCAGGCGCGGAACCTCCGCAATACAC
GAAAGTGTGACGGGGTTACCCAAAGTGTTC
AATAGAACTGTGGTAGGTGAGTAATGTCCCCTAC
TAGAAAGGAGAGGGCAAGGCTGGTGCCAGCCGCCGC
GGTAAAACCAGCTCTTCAAGTGGTCGGGATAAT
TATTGGGCTTAAAGTGTCCGTAGCCGGTTTAGTA
AGTTCCTGGTAAAATCGGGTAGCTTAACTATCTGCAT
GCTAGGAATACTGCTATACTAGAGGGCGGGAGAG
GTCTGAGGTACTACAGGGGTAGGGGTGAAATCTTA
TAATCCTTGTAGGACCACCAGTGGCGAAGGCGT
```

CAGACTGGAACGCGCCTGACGGTGAGGGACGAAAGC
CAGGGGAGCGAACCGGATTAGATACCCGGG
TAGTCCTGGCCGTAAACGATGCATACTAGGTGATGG
TACGGCCATGAGCTGTATCAGTGCCGTAGGGAAACCG
TTAAGTGTGCCGCCTGGGAAGTACGGTCGCAAGGC
TAAAACTTAA

>DQ521781.1 Uncultured archaeon clone SMI1-
GC205-Arc38 16S ribosomal RNA gene, partial
sequence

GGAGGTCACTGCTATTGGGATTCGACTAAGCCATGC
GAGTCGAGAGGGTTCGGCCCTCGGCGAACTGCTCAG
TAACACGTGGATAATTTGCCCTTAGGTGGAGGATA
ACCTCGGGAAACTGAGAATAATACTCCATAGATC
TAGGATGCTGGAATGCACTTAGATTAAAAGCTCCG
GCGCCTAAGGATAAGTCTGCGGACTATCAGGTTG
TAGTCAGGGTAAAGACCTGACTAGCCTACAACG
GATACGGGTTGTGAGAGCAATAACCCGGAGAC
GATATCTGAGACACGATATCGGGCCCTACGGGGCG
CAGCAGGCGCGAAACCTTCGCACTGTGCGAAAGCGC
GATGAGGGGATCCCAAGTGCTTGCTCGTAAGAGTA
AGCTGTTTTTATGTCTAAAAAGCATAGAGAGTA
AGGGCTGGGTAAGACGGGTGCCAGCCGCCGCGGTA
ATACCTGCAGCTCAAGTGGTGATCATTATTATT
GGGCCTAAAACGTCCGTAGCCGGTTTGGTAAAT
GCCTGGGTAAATCGTGTAGCTTAACTATACGAATTC
CGGGTAGACTGCCAAACTTGAGACCGGGAGAGGC
TAGAGGTACTCCTGGGGTAGAGGTGAAATTCTGTA
ATCCTAGGGGGACCACCAGTGGCGAAAGGCGTC
TAGCTAGAACGGGTCTGACGGTCAGGGAC
GAAGCCCTGGGGCGCGAACCGGATTAGATACCCGGG
TAGTCCAGGGTGTAAACGCTGCTTGCTTGATGT

```
TAGTTGGGCTTCGAGCCCAATTAGTGTCGGAGAGA
AGTTGTTAAGCAAGCTGCCTGGGAAGTACGGTCG
CAAGACTGAAACTTAAAGGAATTGGCGGGGGAGCA
CAGCAACGGGTGGAGCGTGCGGTTTAATTG
GATTCAACGCCGGAAAACTCACCGGAGGCGACGGT
TACATGAAGGCCAGGCTGATGACCTTGCCTGATTT
TCCGAGAGGTGGTGCATGGCCGCCGTCAGTTCGTAC
CGTAAGGCGTTCTGTTAAGTCAGATAACGAACGAGA
CCCTCATCTTTAATTGCTACTAGTAAGTCCGCT
TACTGGGCACATTAGAGAGACCGCTGGCGATAAGT
CAGAGGAAGGCGGGGGCAACGGTAGGTCAGTAT
GCCCTGAATCCTCCGGGCTACACGCGCGCTA
CAAAGGCTAGGACAATGAGTTTCAACACCGAGAGGT
GAAGGTAATCTCGAAACCTAGTTATAGTTCGGATT
GAGGGTTGAAACTCACCCTCATGAAGCTGGAATCCG
TAGTAATCGCAGATCAAAATCCTGCGGTGAATAT
GCCCCTGCTCCTTGCACACACCGCCCGTCAAACCAT
GCGAGTTGGGTTTGAGTGAGGATGTAGTTTTTGC
TACGTTCAAACTTAGGCTTAGTAAGCGGGGTT
```

It's hardly the stuff of a Michael Bay movie, but to me it was thrilling. The sequences above represent 16S rRNA genes from two different microbes that were retrieved from our submersible work ~900 meters deep in the Gulf of Mexico. To "read" these sequences, we use a computer algorithm to line them up against genes from a database, such as the one from the National Center for Biotechnology Information (NCBI), and determine which genes in the database are most closely related to them. You can do this yourself right now if you'd like. Type "NCBI blast" into your internet search engine, choose the top hit, choose Nucleotide BLAST or blastn, copy one or both of the entire gene sequences above, paste them under Enter Query

Sequence, and press the button that says BLAST, which stands for Basic Local Alignment Search Tool. Voilà, you now have a list of the genes in the database that are most closely related to these two genes. The top hit should be these sequences exactly since I have already submitted them to the database. After that, I'm guessing you'll get a bunch of 16S rRNA gene sequences that other researchers have pulled from various oceans around the world—you just have to click on them to get their descriptions. It would be difficult to overstate the extent to which these new tools have democratized DNA information: anyone, anywhere, with very little training can find any DNA sequence that exists in the public databases and discover how closely related it is to other sequences around the globe.

The first time I ran a BLAST on the sequences above, the computer returned a list that didn't contain any known microbes at all. The closest relatives were DNA sequences that some other researcher had pulled out of a different part of the deep sea.* The next DNA sequence I tried gave the same useless result. And the next one as well. Almost every strand of DNA we pulled out of these deep-sea samples bore no resemblance to any known organism—just other DNA strands from other environments. My laboratory notebooks are full of handwritten lists of identifiers of these random bits of DNA that contain no meaningful biological information.

At some point, I started to wonder if these DNA sequences belonged to real organisms at all. Maybe I was just looking at a bunch of DNA that was floating around, left over from dead cells

---

* At the time of writing, this is still true, but maybe by the time you try to run this BLAST, the NCBI database will be populated by pure cultures, and you will have real organismal names pop up in the search results.

whose DNA had degraded so much that it was unrecognizable. DNA can, in fact, persist for a surprisingly long time when it's out naked in the environment. For instance, even if you pull up a bucket of seawater that has no fish in it, you can still sequence bits of DNA from fish that have recently passed through that patch of water. This is good news for people tracking fish populations, but bad news for people using DNA sequences to learn about living organisms that aren't visible to the human eye. I worried I was chasing ghosts.

Here's the catch-22 about wanting to be a scientist who makes a monumental discovery. If you're the only one who sees it, you're probably wrong. It's likely a meaningless blip in the data. If you're truly onto something revolutionary, then it's probably so widespread that other people have seen it too. So, while it's fun to celebrate individual scientific heroes, often the biggest discoveries are, by their nature, discovered en masse. Luckily, my particular "discovery" in graduate school was of the group effort variety. Which means it was real.

While I was toiling away on my computer, hundreds of other scientists all over the world were doing the same thing, although I hope at least some of their offices had windows. When I examined the closest hits from my NCBI BLAST searches, I realized that the order of nine hundred As, Gs, Cs, and Ts that I found in Gulf of Mexico sediments matched very closely to the sequences that people were finding in South China Sea sediments, British salt marshes, and Japanese deep-sea trenches. The level of similarity between our sequences was far too high to occur by mere chance. My DNA sequences were therefore not useless molecular junk. I, along with my colleagues around the world, were discovering DNA from living organisms that were on deep lineages on the tree of life that no one had ever imagined existed.

We Were *Wrong* about the Tree of Life

To put our work into context, let's talk about what we knew about the tree of life at the time that we were making these discoveries. In the 1970s, life was categorized into five kingdoms—Animalia, Protista, Plantae, Fungi, and Monera, codified largely through the work of scientist Lynn Margulis (1938–2011).[1] In this schema, archaea and bacteria were lumped together as Monera, as the two were thought to compose a single branch that was equivalent to each of the other branches in terms of its distinctiveness. This five-kingdom organization was adopted by most biology textbooks and became the dominant view.

In the late 1980s, Carl Woese (1928–2012) and his colleagues realized something was wrong with the five-kingdom organization.[2] They had sequenced a gene from what were then called methanogenic bacteria (now called methanogenic archaea or methanogens). They used the 16S rRNA gene, which, incidentally, led to my colleagues and I choosing to use it in our own work a few decades later. When they lined the sequence up against those from "normal" bacteria, like *E. coli*, they found shockingly few similarities. These two organisms were supposed to share a kingdom, but they were no more closely related to each other than they were to humans, plants, or fungi. These archaea were something new entirely. So Woese gave them their own branch on the tree of life and called it a domain, which is the name he and his colleagues gave to a taxonomic level above kingdom. Then he put the bacteria into another branch and called it a second domain. Finally, he squished the other four kingdoms—essentially all visible life on Earth—into a single third domain called the eukaryotes. As you might guess, this did not sit well with the rest of the biologists.

Many biologists were extremely displeased by this microbial takeover of the tree of life. In 1998, the biologist Ernst Mayr (1904–2005) spoke for many of his colleagues when he expressed the following objection to Woese's classifications:

> The eukaryote genome is larger than the prokaryote [archaea and bacteria] genome by several orders of magnitude. . . . This includes not only the genetic program for the nucleus and mitosis, but the capacity for sexual reproduction, meiosis, and the ability to produce the wonderful organic diversity represented by jellyfish, butterflies, dinosaurs, hummingbirds, yeasts, giant kelp, and giant sequoias. To sweep all this under the rug and claim that the difference between two kinds of bacteria is of the same weight as the difference between prokaryotes and the extraordinary world of the eukaryotes strikes me as incomprehensible.[3]

He had a point. It is hard to look at two little dots under the microscope and believe that they have less in common with each other than a mushroom and a baboon, or a tree and a shark, or a slime mold and my uncle, all of whom are eukaryotes.

But we now know that Mayr was missing something. His problem was that he was not properly considering time. A bacterial cell and an archaeal cell may look similar under a microscope, but they've had billions of years to differentiate from each other. Most multicellular life on the other hand, the life that Mayr was so convinced was the pinnacle of evolutionary diversification, has only been evolving for half a billion years. That's enough time to sprout a hundred new legs, grow wings, or learn to shoot your guts through your mouth, grab your dinner with it, and suck all that slop back inside you (as Sipunculids do), but these are just cosmetic changes in the grand scheme of things. It's a much bigger evolutionary leap to go from breathing oxygen, which 100 percent of multicellular eukaryotes do, to respiring the wide array of chemicals that bacteria and archaea do, as changes in respiration require changes in the basic properties that life needs to function.

The mere fact that archaea and bacteria had a three-billion-year head start on evolution relative to animals is, by itself, sufficient reason to assume they carry the majority of diversity on Earth. Nevertheless, microbial dominance of the diversity of life was a tough fact to accept in 1998. As Mayr would proceed to write, "So far, I believe that only about 175 different [archaea] have been described. It is quite likely that further research will find thousands, but hardly more than that. Approximately 10,000 [bacteria] have been named. The number of species of eukaryotes probably exceeds 30 million; in other words, it is greater by several orders of magnitude."[4]

I feel kind of bad dunking on Ernst Mayr. He didn't have the benefit of the DNA revolution that we have now, and he's no longer around to defend himself. If I were in his position in 1998, instead of being a nineteen-year-old college student drinking Guinness by the bucketload while studying abroad in Ireland, I probably would have drawn the same conclusions. But with the full power of DNA sequencing available to us, rather than the small whiff of it that Woese possessed, we know for certain that the number of species of bacteria and archaea are at least the tens of millions that Mayr purports they would never reach, and they are almost certainly orders of magnitude more than that. The number of bacteria and archaea now dwarfs the number of eukaryotes.

Of course, to properly compare the two, we would need some definition of a "species" that is comparable across the domains, which we lack. Currently, a "species" of plant or animal is defined as a group of organisms that doesn't sexually reproduce with other groups. Even though lions and tigers *can* produce adorable cubs together, they don't mate in nature, so they are different species. Unfortunately, this definition is meaningless for archaea, bacteria, protists, and fungi, which primarily reproduce asexually—meaning they reproduce alone.* In microbiology, on the other hand, species are defined by sharing functionalities or genetic features. Usually, the whole genomes must be more than 70 percent similar, or the 16S rRNA genes must be more than 97 percent similar, to be called the same species. If you applied this same definition to animals, the number of animal species would shrink to an even smaller handful because much of animal DNA is pretty similar. All of us in the animal branch of the tree of life breathe oxygen and eat organic carbon, so our genomes are not nearly as diverse as those of bacteria and archaea. Given the necessary differences between the definition of a species in the microbial world and the animal and plant worlds, an apples-to-apples comparison is currently difficult.

One alternative way to determine how many different organisms are in each of the three domains would be to look at higher taxonomic groupings, like phyla. The taxonomic hierarchy is domain, kingdom, phylum, class, order, family, genus, species (every time I write it, I have to whisper, "King Phillip Came Over From Great Spain," otherwise I get the order wrong). Luckily, the three-domain theory didn't change the taxonomic hierarchy; it just added "domain" at a higher level than "kingdom." This hierarchy helps us to talk about how long-ago

---

*Actually, many plants and animals reproduce asexually too. Snakes and lizards often do this. There's an aphid, which is a type of insect that lives on common garden plants, that is so good at asexual reproduction it can be born pregnant! (Moran, Nancy A. The evolution of aphid life cycles. *Annu. Rev. Entomol.* **37**, 321–48 [1992].)

familial lineages diverged from each other. Two *species* may have diverged recently—even in the past hundred years or so. Two *phyla*, on the other hand, may have diverged billions of years ago. Since longer time periods allow for more evolutionary changes, two species in the same genus should be more similar to each other than two species that belong to different phyla. So what if we scoot around the issue of comparing the number of species across the three domains by aiming higher? What if we compare the number of phyla instead? Phyla are even less well-defined than species, but since they cover more evolutionary time, they are much more consequential for biological diversity.

There are at least 27 known phyla of archaea[5] and 95 of bacteria,[6] although at the rapid rate of DNA sequencing, these numbers frequently increase. If you use a new metric for phylogeny being applied by Phil Hugenholtz and his colleagues in Australia, there are 18 phyla of archaea and 148 phyla of bacteria. In contrast, the eukaryotes have 75 phyla, although only two of them are the plants, animals, and fungi that so entranced Mayr; the rest are the single-celled organisms formerly called Protista in the five-kingdom schema.[7]

Based on this new information, we can conclude with confidence that Ernst Mayr's eyes deceived him; the microbes, particularly the archaea and bacteria, have a much greater range of evolutionary and functional diversity than eukaryotes do. Those ~3.8 billion years were not wasted time for life on Earth. Life was busy evolving. And much of that evolution was happening in the intraterrestrial lineages that I, as a graduate student, was trying to disentangle.

What we found was astonishing: these DNA sequences were new phyla. Just from one submersible dive, we had discovered phyla that had yet to be described.* If it was that easy to find these, then there were certainly more to follow. We suddenly realized that Woese's discovery was the tip of the iceberg. The bacteria and archaea contain many more deeply branching lineages than he was able to see with the technology available at the time, and some of them didn't even fit cleanly into any of the three domains at all. They were so weird that they threatened to upend Woese's new three-domain tree of life that had only gained widespread acceptance a few years beforehand.

Recently, I was revising something for publication, and I noticed that I had written, "Subsurface microbes are different than *humans and other plants.*" This is a little embarrassing, but not completely unexpected. When you spend as much time as I do combing through the evolutionary relationships of all living things on Earth, you're constantly reminded that relative to the vastness of microbial diversity, humans and plants are not very different from each other. This method of direct DNA sequencing from environmental samples is so powerful that it has made discovering new species as mundane as cooking dinner or washing your car. I could do it daily if I could produce DNA sequences fast enough. The trick is that most of the diversity of life on Earth is microbial. Since we can't see most of them, we've managed to coexist with them without even knowing they were there. It's like we we've been living beside magical

---

* These phyla have now been named the Thermoplasmatota and Lokiarchaeota, and I want to emphasize that we were not the only ones discovering them—they were popping up all over the world. In chapter 9, I will discuss how Lokiarchaeota change our view of early life on Earth.

unicorns throughout human existence—they were just tiny, single-celled, and far, far more interesting than horned horses.

What Do You Do after You Catch a Unicorn?

These discoveries were a stunning thing to be a part of, but, if I'm being honest, the whole thing was a little anticlimactic. It's like finding out you've discovered an alien because you get a printout saying, "This is an alien." If our DNA sequences had matched to known organisms, we could have made educated guesses about what their lives were like in the subsurface. But as it was, we were totally in the dark: no pictures, no cool space-ships, no telepathy. Disappointing, right?

We needed to learn about these newly discovered beings, but we couldn't just pull out binoculars and observe them hunting prey or munching leaves. We couldn't conduct experiments on them as laboratory cultures because, as I've mentioned, they had somehow evaded all culturing attempts thus far. What we needed was to do something really big. We needed to invent a new way to study their physiology.

In our modern scientific method, developed by Francis Bacon (1561–1626), all variables in an experimental system need to be tightly controlled so that they can be changed one by one to test hypotheses. But unfortunately, this doesn't work very well for the intraterrestrials. Because so few of them yield to laboratory conditions, we're forced to study them on their own terms, in their natural environment, where things are messy and chaotic. No one in my undergraduate classes taught me how to do science when you can't control anything at all. To find out what these intraterrestrials are like, we have to step *way* outside the scientific comfort zone. I mean, we don't even get to control what organisms we're experimenting on! Luckily, there are some fantastic new techniques that allow us to glean maximum information from experimental conditions so uncontrolled they would probably make Francis Bacon blush. And an inevitable part of forging a new path is making mistakes.

In November 2006, I took a small skiff with my friend and colleague, Howard Mendlovitz, to the White Oak River estuary, close to where I grew up, so I could jump into the shoulder-high water and take push cores of mud for my PhD project. If I had just wanted DNA sequences from intraterrestrials, I could've grabbed a small scoop of mud, which is easy, and taken it back to my lab. But I wanted more than DNA sequences. I wanted to learn about what these organisms were actually doing out in nature. For that, I needed to collect many three-foot-long cores, so I could look for subtle impressions the intraterrestrials made on the layered chemistry around them so I could test hypotheses about what they might be doing. I had to dig these cores out by hand—no submersibles or oceangoing vessels were at my disposal for *this* project.

This field site was so familiar to me that I couldn't imagine myself being in any danger. I was a short swim from where my dad used to live when I was a child. It was a sunny fall day, but even though I put on a wetsuit for warmth, I was cold the moment I jumped in the water. Howard suggested that I abort the mission and go get warm, but I waved him off. Having always aspired to stoicism, I pulled up core after core until Howard leaned over the side of the boat and looked me in the eyes. He pointed out that I was no longer being effective in my work. I was tugging on a core that I had shoved into the sediments, but I was making no progress unsticking it from the muck. So I climbed on board and marveled at my thigh, which was shivering independently of the rest of my body.

Howard looked me over and said, "I think we better get you back to land. Tape a cap onto this last core and we'll go."

I stretched out electrical tape to the sopping wet core barrel, but I couldn't remember how to put it on. Howard looked at me like I was nuts and said, "Karen, you can't put tape on wet plastic— dry the core off first, like we've been doing for all the other cores."

By this point, my whole body was convulsing, so Howard decided to speed up our departure. He said, "Fill up this bucket with water, and splash it over the boat so we can clean the mud off and go." I distinctly remember holding the empty bucket by the handle, and gingerly placing its butt-end on the surface of the water, without bothering to tilt it on its side. I stared at it, wondering why the water wasn't going inside. That's when Howard said, "Don't worry about cleaning off the deck, just put warm clothes on right now!" I stripped down to my bathing suit and looked at my eggshell-white feet. They were curling up in some sort of pre–rigor mortis state, so I tried to shove them into my shoes. For the last time before we finally got on our way back to shore, Howard looked over and shouted, "Karen, *pants go on before shoes!*"

My stubbornness and lack of respect for the serious nature of the situation had brought me to the early stages of hypothermia. One moment, I was telling myself, "Okay, suck it up, buttercup, you're just a little chilly," and the next moment, I was in real danger. Thanks to Howard, I made it back to the shore just fine and warmed up in the heated truck. My poor choices that day affected the science that came out of this work—the cores were too short to be useful. They appear in my PhD dissertation but didn't add enough in the way of new information to be included in the peer-reviewed publication. Now, when I'm in the field—whether I'm descending into the deep ocean or lazily splashing in a sun-dappled pond—I run through all the possible ways my situation might change and try to make decisions that minimize risks. No sample is worth my life.

I returned to the White Oak River estuary many times over the course of my PhD, with a lot less drama. However, I made further mistakes when I analyzed the data. I spent my whole PhD searching for one type of intraterrestrial, a methane-metabolizing microbe not too distantly related to known organisms. But what gradually became clear to me was that I was barking up the wrong tree. The dominant organism in these mud samples was something even weirder than the ones I had been looking for: it was a novel phylum that had only ever been found in a boreal forest lake,[8] oceanic sediments,[9] and deep South African gold mines.[10] Fumio Inagaki, of the Japan Agency for Marine-Earth Science and Technology (JAMSTEC), named the phylum the Miscellaneous Crenarchaeotal Group (MCG), like they were errant file folders. Even though I wasn't anywhere exotic, just in an estuary with normal pressure, temperature, and pH, this otherworldly MCG group was the most abundant thing there. I had been chasing the wrong intraterrestrial. The ones I wanted were the MCGs.

What we needed was more information—anything—to tell us about what these MCGs' lives were like. When I finished my PhD, I was deciding where to go for a postdoc* when Bo Barker Jørgensen at Aarhus University in Denmark, said, "I think these MCG are fascinating too. Why don't you move to Denmark, work with me, and we'll try to sequence their whole genomes?" We agreed that our mark of success would be getting *one* gene besides 16S rRNA that we could clearly identify as belonging to these strange organisms. This was easier said than done, since

---

* In my field of science, it is common to spend a few years after your PhD doing a postdoc, where you focus on research and apply for positions as a permanent scientist.

the process of DNA sequencing erases information about which gene came from which organism.

Our plan was to physically separate a single cell out of the muck so when we sequenced its DNA, we knew all the genes came from that one organism. But this technology had never been used before in muddy samples. And separating tiny cells from equally tiny bits of mud was excruciating. When I moved to Denmark, my first task was to help Bo buy a newly invented laser microdissection microscope, so we could cut a tiny cell off a filter, flick it into a tube with the laser, and sequence its genome.

Technically, my first task was to learn how to say, "Where is the delivery room?" in Danish, because I was five months pregnant. And, as luck would have it, the other postdoc I was working with on this project, Dorthe Petersen, was also pregnant. Because of the ticking clocks in our bellies, Dorthe and I worked like fiends to cut out tiny cells with the laser microscope, together with our not-pregnant colleague Lars Schreiber. But our babies beat the technology. By the time they were born, we hadn't yet figured out how to prevent static electricity from whisking the cells away after we cut them out with the laser.

With Denmark's generous maternity leave, my colleagues and I had plenty of time to rejigger our approach. We had known that there was another new technology being pioneered by Ramunas Stepanauskas at the Bigelow Laboratory for Ocean Sciences in Maine that produced single-cell amplified genomes by the boatload. The only catch was that Ramunas was working in seawater, and the very expensive machine he was using to do this, called a fluorescence-activated cell-sorter, would be destroyed by getting any little pieces of dirt inside it. And our samples were *all* dirt. Nonetheless, single-cell sorting seemed like something worth exploring, so Dorthe and I left our year-old babies in the capable hands of our husbands and flew to

Maine to talk to Ramunas. He immediately saw the importance of using his technique for exploring strange life in Earth's sub-surface. After some consideration he said, "What the hell, let's try it." We sent him samples we had dredged out of the muck at the bottom of Aarhus Bay, Denmark.*

Not more than a month later, Ramunas sent us tens of little vials, each of which contained DNA amplified from a single cell. It was the stuff of my dreams. We sequenced each genome as quickly as possible, and Lars forged a pathway to analyze the genomic data, since this had never been done before with these types of organisms.

Combing through the list of genes felt like reading ancient scroll, except that ancient scrolls are at most only a few thou-sand years old. These genes were from lineages that were *billions* of years old, and we were the first people to ever lay eyes on them. Bo and I had knocked our "just *one* gene" plan out of the park. We now had thousands of them.

The first thing we noticed was that most of these genes were so weird, we couldn't determine anything useful about them. Even today, the vast majority of this "genomic dark matter" re-mains completely uncharacterized. It's present in smaller amounts in humans and *E. coli* too. So, as enticing as that stuff was, we had to leave it for a future discovery.

What we did find was something truly unexpected for next-to-dead intraterrestrials living off old organic matter in marine sediments. One of their genes was for the protein gingipain, which is an enzyme secreted by pathogens to dissolve gum tis-sue, causing the disease gingivitis. But these intraterrestrials are

---

* I figured if this type of microbe was abundant in coastal North Carolina, it was abundant in coastal Denmark too, and I was right.

never found in our mouths, so why would they have an enzyme like this?

Perhaps even weirder was the idea that these cells would be releasing enzymes outside of their cells at all. Enzymes are expensive to make, and these cells were starving. Even if the cells were using these enzymes to break down food, which is probably what they were doing with them, secreting them is like leaving $100 bills all over the city in the hopes that someone picks them up and buys you a sandwich. It's an inefficient way to find sustenance. I could've overlooked these genes if there were just a few of them, but each cell had many copies of them. And gingipain was just a drop in the bucket. One genome alone had seven genes for gingipain and five more copies for similar enzymes!

We sent our finding to the journal *Nature*, but they needed more evidence that these genes were doing what we thought they were doing; perhaps by designing a new assay for these enzymes to show they are active in this environment. Lamentably, I had no idea how to do this. That night at dinner, I was complaining about this to my husband, Drew Steen, also a professor at the University of Tennessee, when he looked at me incredulously and said, "You *know* what I do for a living, right? I design assays for new enzymes in messy environmental samples."

Luckily, he forgave my obtuseness and designed a new assay for the enzymes whose genes we had found, and I gave him some of our frozen mud so he could try it out. A few weeks later, he called me while I was at a conference abroad. I was worried it was a problem with our daughter, but he said, "Not only is gingipain active in your samples, it's *six* times more active than the types of enzymes we would normally look for in sediments like these." We had discovered a new function for these intraterrestrials, and we had also discovered a new carbon-degradation pathway for marine sediments in general. We published the paper[11] and it launched much of the research I've been doing ever since.

Right about the time we were producing this work, a much greater seismic shift was brewing. Jillian Banfield and Katrina Edwards, who had previously worked together at the University of California Berkeley discovering archaea that grow[12] at pH 0 (I'll talk more about how archaea achieve this amazing feat later in the book), were busy paving the way for subsurface biology to be a real, mature field of science, rather than just a collection of one-off projects. Katrina moved to the University of Southern California, founded the National Science Foundation (NSF)–funded Center for Dark Energy Biosphere Investigations, and launched multiple deep-ocean drilling projects. Through these efforts, she and her colleagues brought funding and access to subsurface biosphere samples to researchers such as myself and others from all around the globe. At the same time, Jill was throwing bioinformatics into high gear, accelerating our ability to use DNA to interpret microbial physiology. Years earlier, she and Gene Tyson obtained the first genomes directly from an environmental sample,[13] but I never considered using their methods because I thought my microbial communities were too diverse. Right when we were publishing our small handful of MCG genomes, Jill and Kelly Wrighton released an astounding forty-nine genomes from a deep subsurface aquifer,[14] and the numbers have exploded into the thousands since then. Over the past ten years, Jill, Gene, Kelly, and their colleagues have created a rich garden of analytical tools and have used them to discover a large swath of the intraterrestrial world. From their work, we can now combine the ability to know *what* an organism is with a deep understanding of its potential physiological capabilities—all without having to grow it in culture. In this way, Jill Banfield and Katrina Edwards enabled subsurface biology to become a mature field in its own right. Because of the efforts of these two women, we can now be full-time subsurface geomicrobiologists, rather than "normal" surface biologists who occasionally dabble in this strange subsurface place.

Building on some of the techniques developed in the Banfield lab, Fengping Wang at Jiao Tong University, China, produced much higher-quality genomes from that MCG group that I had been chasing and gave them a much nicer name: Bathyarchaeota. *Bathy* means "deep" in Greek, and she and her colleagues chose it because these organisms are deeply branching on the tree of life, and they live deep inside Earth's subsurface.[15] Next, the team, along with Kai-Uwe Hinrichs at the University of Bremen, Germany, took this research another huge step forward. They fed Bathyarchaeota one of the toughest foods on the planet—tree bark—to see if the cells could eat it. But finding out if Bathyarchaeota eat tree bark is difficult because Bathyarchaeota only exist in natural samples that contain billions of species, so it's hard to know who is eating what. So, my colleagues did a neat thing that allowed them to figure out which organisms were actually eating the tree bark. They chemically altered the individual carbon atoms of this tree bark chemical, technically called lignin, mixed it with estuary sediments, and waited nearly a year to see if the Bathyarchaeota would take the bait.

At the end of the experiment, they found that Bathyarchaeota's biomass did in fact contain the chemically altered carbon atoms that identified them as having come from the lignin. This means that Bathyarchaeota, this strange deep branch on the tree of life that has been sitting in muck that now contains oysters and clams since before oysters and clams evolved, eats chemicals from tree bark—or at least the types of Bathyarchaeota that we see in estuaries do. We don't know yet about the deep-sea versions, which would have had far less frequent contact with tree bark. Since trees only evolved about half a billion years ago, we don't know if this is a new adaptation that arose when trees appeared or if Bathyarchaeota have always eaten things that chemically look like lignin, so when trees showed up, they merged gracefully into Bathyarchaeota's preferred palate.

Bathyarchaeota turn out to be more than just a nerdy oddity. They are also abundant in the wastewater and sludge reactors that remediate our human and agricultural wastes. So, these Bathyarchaeota may directly impact our own lives. We're just getting started finding out what they can do.

We have many more tricks up our sleeves when it comes to studying the Bathyarchaeota and all the other intraterrestrials, with these technological innovations born of necessity. We can put little chemical tags on cells that tell us who they are, what they're eating, how fast they're growing, and whom they like to cuddle up to. And sometimes, if the samples are behaving, we can use high-powered microscopes to see what individual cells are consuming, even out in messy nature. In this way, we are piecing together information about our strange fellow Earth-lings, atom by atom.

But not all intraterrestrials are so difficult to work with. Some of them grow quite nicely in laboratories. But this doesn't mean they're "normal." These types of intraterrestrials are called ex-tremophiles, and they grow under conditions that would kill every run-of-the-mill organism on this planet. They push so close to the boundaries of what it means to be alive that they've altered our assumptions about where those boundaries even lie.

5

HOW TO LIVE INSIDE
A VOLCANO

EVEN THOUGH we are surrounded by a wondrous array of life, ranging from snakes and beetles to whales and people, the intraterrestrials have shown us that this is just a narrow slice of life on Earth. What if the *conditions* that can support life are also broader than we imagined? There are places on this Earth with conditions that would instantly kill most of the plants and animals that we know and love. But to the intraterrestrials we call extremophiles, these places are home. And perhaps because dunking your culture bottles in a vat of boiling water kills the microbial weeds that you don't want, many of these extremophiles actually grow well in laboratory culture. This has put us in a powerful position to figure out how life deals with extremes in temperature, acid, pressure, and salt.

In this chapter, I'll describe some of the discoveries folks have made about how life does these amazing feats. But to study extremophiles, sometimes you have to try to *become one yourself*, which is not easy to do. As we will see, microbes have had the leisure of ~3.8 billion years to hone the art of feeling comfy in a violent volcanic crater. I, on the other hand, had to get my act together in less than twenty-four hours.

Descending into an Active Volcano

Poás volcano, located about fifty kilometers from San José, the capital of Costa Rica, is a large stratovolcano,* the product of many eruptions over geological time. Instead of coming to a pointy top, it has a massive, curved bowl, which stretches a mile across and a thousand feet deep. It looks like a giant walked by and scooped out the mountaintop like it was a scrumptious spoonful of sorbet. Unfortunately, the real taste and smell of this environment is considerably less appealing. The heat of the volcano boils elemental sulfur, which is normally a chalky solid, into a delicate pudding, such that pure, yellow, goopy sulfur constantly gurgles out of the crater floor. At the crater's center is a lake that is an eye-popping shade of turquoise. It gets its color from the sulfur globules and clay minerals that churn through its waters.

Poás's crater lake is so beautiful that Costa Rica has made it a national park with educational displays and an overlook offering a gorgeous view from a safe distance.† However, despite the allure of the Instagram-worthy lake, no one is allowed to hike down into the crater itself, even when the volcano is calm. The reason for this is that the whole thing might erupt, possibly without warning. And when Poás erupts, it doesn't

---

* This is a type of volcano that's been erupting episodically for long enough to build up a big pile of alternating layers of lava and ash.

† If you're so inclined, take a break from reading and type it into Google Maps. Make sure you turn on the satellite view in the lower lefthand corner. You can see the lake and multiple eruptive events in the denuded crater. Just to the southeast of the current crater is an old crater, now called Laguna Botos, which hasn't been volcanically active in modern times.

belch rivers of lava like some other volcanoes. No, it's much more dangerous. Poás is what is called a phreatic volcano, which has a deep magma chamber, well away from sight, and when water trickles down into it, pressure builds up dangerously deep underneath the crater, like a mountain-sized pressure cooker. When the pressure reaches a breaking point, the whole mountain top explodes in an instant. Any humans at ground zero would be shot hundreds of meters into the air with the force of a massive bomb. When they fall back down and smack into the ground, they'll get hit by the boulders that were also ejected along with them. This violent ride will happen over and over again—keeping a person's body aloft like a Bingo ball in a hopper. An eruption of sufficient magnitude would mean certain death for someone hiking in the crater. If you're at the viewing overlook at the visitor center when it starts erupting, at least you can outrun the deadly rain of boulders.

But eruptions aren't the only reason that people are not allowed to hike down into the crater. Even on a calm day, with the deep waters simmering peacefully underground, the rocks will try to kill you. This is because these are not normal hiking rocks like sturdy granites or weathered river stones. The rocks in Poás are basaltic glass, shattered and strewn haphazardly from whatever eruption happened most recently. If you trip and fall, the ground itself might slice your hand open. The rocks are also unstable, scrabbly, and positioned at odd angles—a nightmare for anyone with weak ankles or low cardiovascular fitness. And there's always a chance of earthquakes. A strong earthquake while you're in an up-to-code building is one thing, but when you're at the bottom of a gigantic funnel of sharp glass, even a small earthquake could trigger a deadly landslide. The only good news is that you won't get lost on this hike. Even though the iron-rich volcanic rocks *could* provide nutrition for growing trees, the frequent eruptions force them to cower at the crater's edges like nervous aunties fretting over a misbehaving child. Therefore, any hiker in the crater would have a clear line of sight to the safety of the visitor center. Finally, if you make it to the crater's crown jewel, the ethereal lake the color of elves' sweetest dreams, the lake itself might kill you. It's made of the same acid found in batteries, so if you slip and fall in, you probably won't make it out alive.

What living thing could possibly enjoy this fuming lake of sulfuric acid that explodes every now and then? Given the difficulty of getting samples, we are not awash in data to answer this question. But what we *do* know gives us the implausible answer that this volcano might have more life to offer than meets the eye. Poás is world-renowned for its thick sulfur deposits that are just itching to react with oxygen to release energy. However, the reaction is sluggish enough that the sulfur just kind of sits there for a while. This chemical environment is like money in the bank for microbes, as it provides abundant fuel for chemolithoautotrophy, so long as they can survive the unstable, highly acidic surroundings. Microbes living in the crater lake are like real estate developers who build beach houses right on the sand dunes of the hurricane-pummeled coastline of North Carolina. There's a great chance their investment will be flattened, but they gamble that they'll get many years of rental income before that happens. High risk, high reward. A similar temptation presents itself to a curious scientist who loves a challenge. If you want to know just how microbes cling to this unstable, hostile land, you have to put your skin in the game too. Despite all the reasons not to go, you have to hike down to the crater yourself and pick up the samples by hand.

This is just what I did in February 2017. Our team, which consisted of an unwieldy group of twenty-five scientists plus two separate film crews documenting our work, was spending two weeks driving around sampling hot springs all over northern Costa Rica. So far in the trip, we had weathered some dicey interactions with hot springs along the way and had made some arduous hikes through thick jungles to reach springs on the flanks of volcanoes. But nothing matched the potential danger of hiking into Poás. Our colleagues at Observatorio Vulcanológico y Sismológico de Costa Rica (OVSICORI)—who take turns hiking down into Poás's crater regularly to collect samples and monitor the threat level that the volcano poses to Costa Rica's nearby capital city—were the ultimate authorities on who would be allowed into the crater. To limit the number of people who would be hurt if something went wrong, they whittled the crater-bound team down to the smallest number of people required to collect the samples, and only greenlit one of the film crews to go with us. The other film crew seemed relieved to stick to the wide-angle shots from the visitor center.

The night before we hiked into Poás, the outreach coordinator, Katie Pratt, and I were gearing up for the trek and chatting with Steve Turner, a seasoned volcanologist, about what to expect. At some point in our conversation, it became clear that Steve wouldn't be joining us in the crater the following day. Katie and I were baffled as to why a volcanologist on our trip wouldn't be going into the volcano. That's when he said, "Oh, I have to go inside so many volcanoes for my job that I avoid it whenever I can, just to cut down on my exposure to risk." Katie and I looked at each other like deer in headlights. How much risk, exactly, were we taking on?! But it was too late to change our minds now, and neither of us wanted to miss this once-in-a-lifetime opportunity. Plus, the volcano *had* been really quiet for many months, so the chance of a safe journey was high. Nevertheless, the revelation that Steve wasn't going with us flipped some sort of switch in our brains, and our demeanor grew solemn. For some unfathomable reason, Katie and I both started quietly sipping bottles of Gatorade. Did we think that proper hydration could prevent a volcanic eruption? Hard to say.

The morning of our descent into the crater was magical. Blue skies, a quiet volcano, and jittery scientists. Despite the geological hostility, I quite enjoyed the hike down. For the first tens of meters, we pushed through bright green broad-leafed shrubs. After that, it was all bare rock, rubble, and consolidated ash piles with rain-built sluices. I have hiked in the Appalachians, the Alps, the Rockies, the Adirondacks, and the Sierra Nevadas, but none of them were anything like the hike down into the crater of Poás. The descent was like scrambling down a parking lot retaining wall covered in broken beer bottles. Everywhere I looked was a sea of textured or glassy rocks. It had an industrial feel, as if the gravel had been piled up by bulldozers. If someone had told me, "Imagine yourself in nature," this is probably the last landscape that would come to mind. But the stark ground I was navigating was just as natural as a babbling brook in a lush forest.

We finally made it to the bottom of the crater, but we hadn't yet reached the lake, which was obscured by one last mound of basaltic shrapnel. Coming around the final bend, I was shocked by the blinding brightness of the lake, as well as how enormous it was. What had appeared as no more than a pretty blue bauble in the distance when we started hiking that morning had blossomed into a sprawling sea. To give a sense of the scale, our bodies on the shore of the lake were too small for our colleagues back at the visitor center to see. Just then the winds shifted, and I was hit with the full sensorial experience of Poás lake, which is technically called Laguna Caliente. I was immediately reminded of 25 percent strength salad vinegar you can purchase in supermarkets in Germany. If you don't water it down before you use it, your mouth will burn, your eyes will tear up, and if you catch enough of it in your nostrils, you may even gag from the hit of sharp acid. This is what it felt like to stand next to Laguna Caliente. Any interest I had in jumping in for a swim was immediately replaced by a desire to pat my respirator for comfort.

While my attention was entirely focused on the lake, my OVSICORI colleague Maarten de Moor trudged past me on the way to start taking gas samples and said, "Keep your feet moving so your boots don't melt." The ground I was standing on was 100°C. Then he came back and said, "But don't move them around so much that you lose your footing. There's nothing to stop you on this slope and you'll slide into the lake." For a moment, the perilous nature of the situation overwhelmed me: I was terrified, awestruck, and having one of the most transformative experiences of my life. But we had measurements to make and samples to collect. And we were in a hurry.

Along with my colleague Donato Giovannelli, a microbiologist at the University of Naples "Federico II," I inched closer to the water's edge to collect some samples. For being so dangerous, the lake appeared serene, lapping at the edges of the rocks like a small pond. But this illusion of benevolence evaporated when I dipped the pH meter into it. The meter read 0.85. To this day, this remains the most acidic reading I have taken in nature.

Next, I needed to reach into the water with my gloved hand and fill up a giant syringe, so we could push the water through a filter that would catch the microbes—a treacherous endeavor, as the ground at the edge of the battery-acid lake was covered in slippery clay. Luckily, Donato's years of rugby and my years as a cheerleader prepared us for just this moment: Donato bent low and locked arms with me while I leaned out over the lake and filled syringe after syringe with the toxic fluid. The clay deposits at the edge threatened to send my feet flying at any moment, but they held firm. Surprisingly, the water wasn't very hot. It was about 40°C. If it hadn't been full of skin-melting acid, it would indeed have been perfect for a nice soak. We completed the tense sampling without incident and brought the water-filled syringes over to a relatively flat surface, where the ground was sub-boot-melting temperature, so we could start the long process of filtering out the microbes.

Pushing water full of clay and sulfur particles through a filter is no easy task. Normally, turbid water like this would take us forever to filter. Normally, our field guide, Carlos Ramírez, would ignore us while we were doing our work, so he could focus on the logistics of getting to the next site. Normally, it wouldn't matter how long we sat at a site, cursing as we tried to squeeze the last drop of liquid through a clogged filter. But this was not a normal day. We were in the crater of an active volcano, and every extra minute we spent down there increased our chances of disaster. Carlos grabbed one syringe in each hand, walked over to a flat rock and began using it as a base, so he could push the water through the tiny 0.2μm filters by bearing down on the plungers. As we worked, I thought that maybe it was a good time to ask him about our escape plan, in case an eruption did happen. He looked up brightly and said, "Yes, of course! I'm so glad you asked. The exit strategy is simple. If this crater starts erupting, you should turn toward it and enjoy the view . . . because it will be your last."

It may sound like Carlos was being overly dramatic, but fifty-four days after we climbed out of that crater, Poás experienced its largest eruption since the 1960s. Jets of steam, rock, and acid shot kilometers into the air for two weeks straight. The plume was visible all the way from San José. Our colleagues from OVSICORI, who were, gratefully, safe and sound in their offices during the eruption, sent us the last moments that their webcams recorded before they were destroyed in the blast. As I watched these clips in disbelief, my eyes searched for the place where my feet had been nervously shifting on the scalding, unstable ground just a few weeks before, as that same ground exploded into a giant fountain. Steve's decision to avoid the trip and minimize his risk of exposure seemed sensible in retrospect.

When the volcano calmed down a bit, a few of my colleagues ventured up to the long-closed visitor center and sent us eerie videos of the dust-covered viewing platform, enshrouded in a thick foggy smoke. It looked like an alien landscape compared to that clear day two months ago when I swung my legs over the rail of the overlook to begin my hike. When it was finally safe enough to send drones out to check on the lake, my colleagues discovered that it was gone, having vaporized in the eruption. Over the following months, the lake refilled with a mixture of deep volcanic waters and the steady input of rain for which Costa Rica is famous.

How to Withstand Acid and Explosions

My adventure into the volcano lasted only a few hours, yet I felt fortunate to come away unscathed. Microbes, however, spend their entire life chilling down in the acid lake, even though acid kills most microbes as easily as it would kill me.[1] Acid is made by the presence of excess protons in water. But these excess protons are positively itching to react with everything else too, not just water. This is a serious problem for microbes, because almost all the important stuff that makes up a cell (proteins, DNA, RNA, lipids, and other metabolites) contains chemicals that will stop working if they get too many protons stuck on them. DNA can only do its double helix thing because the two sides are held together by hydrogen bonds made of these protons. In acid, each helix will react with the protons in the water instead of the ones from the other helix, unraveling because the "Marcos" never find their "Polos." The same principle disrupts protein, RNA, and lipids too.

As if broken biomolecules weren't enough, acid destroys life on a more fundamental level. Every living cell must create a gradient of electricity and/or pH across its membranes.* They maintain these gradients by keeping careful watch over how many protons are inside and outside of each cell. Acid sprays protons around like heavy artillery, threatening life's precarious proton balancing act.

Here's how the process is supposed to work. All living cells continually push protons outside of their membranes. This buildup of protons outside the cell leads to an imbalance in chemical concentration that drives protons back into the cell to even it out again. This chemical/electrical pressure is called the proton motive force. It's called "motive" because when those protons rush back into the cell, they kick a flywheel of a protein called ATP synthase that makes ATP for us. ATP provides the energy we use whenever we need to make something happen inside our cells, so in a way it's like cash for a cell. We can spend it for whatever we need our cells to do. Without ATP, humans and microbes alike die. And beyond providing ATP, these protons that "fall" back into the cell also power other parts of the cell directly. For instance, the motor proteins (that's not an analogy—they're actually called motor proteins) that crank flagella and allow the cell to swim are powered directly by protons rushing into the cell. Using the proton gradient directly allows flagella to whip around much faster than if they had to

---

*A gradient across a membrane means that there's more of something on one side than the other, so there's a natural "urge" to even out the distribution and move across the membrane.

wait to make and then spend ATP. It's like swiping your credit card rather than going to an ATM and getting cash.

This proton gradient often creates a pH gradient since protons react with water to make acid. Given the importance of this pH gradient to energy production, and thus life, this means that cells actually *want* their outside to be more acidic than the inside. Life usually puts a lot of energy into maintaining this pH gradient so cells can keep their tidy machinery running. This brings up an intriguing possibility. Maybe cells in volcanic acid lakes have it "made in the shade." Maybe they can power themselves for free, drawing off the limitless supply of free protons just dying to rush in and kick all those good protein flywheels and make ATP for free. But that's the problem. If the cell doesn't do anything to stop it, that massive pH gradient will rush in like water through a dam breach. Even if you're thirsty, drinking from a firehose might kill you. While most of our human cells at neutral pH are concerned with pushing protons out of the cell, microbial cells in volcanic acid lakes—a pH of 0.85!—are more concerned with preventing the protons from rushing in all at once like a tidal wave.

Yet when I dropped my pH meter into that deadly lake at Poás volcano, it was bumping into living microbes that were presumably thriving. How could microbes possibly survive in nearly pure acid? They have essentially two options. The first option is to evolve all their biomolecules so that they are "tuned" to pH 0.85 and only function properly when they're bathed in acid. In this world, DNA would be held together by strange new types of molecules that are so resistant to protons that they would fall apart in neutral pH. There is a fascinating subfield of organic chemistry where scientists try to invent new biomolecules that function under such implausible conditions. However, at the time of writing no one has discovered such a

FIGURE 1. Running beside the truck in Argentina after the vicuña have run away. My colleague Carlos Ramírez is in the background. Photograph: Joy Buongiorno. 2019.

FIGURE 2. Students Lovell Smith, Leketha Williams, Gage Coon, and myself doing the laborious work of sectioning a core, June 8, 2023, Morehead City, North Carolina, at the University of North Carolina Institute for Marine Sciences. Photograph: Joy Buongiorno.

FIGURE 3. Boarding the *Alvin* submersible, August 31, 2023,
Astoria Canyon off the coast of Oregon. Photograph: Ellen Lalk.

FIGURE 4. Enshrouded in volcanic fumes, crater of Poás volcano, February 2017, Costa Rica. Photograph: Donato Giovannelli.

FIGURE 5. Happy dance with our first true permafrost samples. Also pictured is Christian Rasmussen. The permafrost drill is yellow. February 2021, Kvadehuken, Svalbard. Photograph: James Bradley.

FIGURE 6. Bouncing down Irruputuncu volcano, Chile, March 2022. Photograph: Jacopo Pasotti.

FIGURE 7. Taking notes on Isluga volcano, Chile, at 6,000 meters, March 2022. Photograph: Jacopo Pasotti.

FIGURE 8. Planning experiments with Donato Giovannelli on the back of a truck in Chile, March 2022. Also pictured (L-R) are Viola Krukenberg, Gerdhard Jessen, Jenny Blamey, Carlos Ramírez, and Agostina Chiodi. Photograph: Jacopo Pasotti.

FIGURE 9. Sampling a deep spring on the border of Argentina, Chile, and Bolivia, March 2023. Photograph: Jacopo Pasotti.

FIGURE 10. Proudly displaying a live horseshoe crab at the North Carolina Aquarium at Pine Knoll Shores, near my hometown of Beaufort, North Carolina, circa 1987. My brother, Finley Lloyd, is in the background.

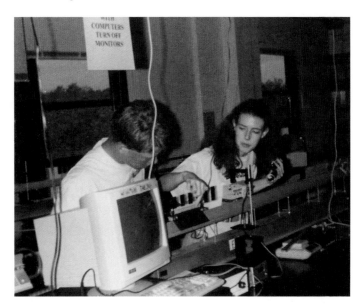

FIGURE 11. Explaining the frictionless air track from my physics lab at the North Carolina School of Science and Mathematics to my brother, Finley Lloyd, 1995.

thing in real life. Microbes that truly like acid do keep a lower pH inside their cells relative to the rest of us,[2] but so far it seems that they don't push it far below about 4.6.

The second option is to battle the acid itself. The first way they fight the acid is by trying to patch any holes that let protons leak into their cells. Acidophiles (microbes that prefer acidic environments) are built like tanks. They use special types of lipids and have very small pore sizes on the proteins that span their membrane, so very few protons can leak into the cell. Acidophiles also keep lots of acid-buffering molecules inside their cells to soak up any protons that do leak through. And by keeping a stockpile of chaperone proteins, they can quickly repair the biomolecules that are broken by the protons that make it inside. If the gambit works, they can keep their internal pH high enough that their biomolecules won't fall apart, and they can maintain the pH and charge gradient across the membrane.

Some acidophiles go to even more extreme lengths to win the proton war against acid. They can invert the charge difference across their membranes. While the rest of life keeps a negative charge inside and a positive charge outside their cells, some acidophiles keep their innards positive. How do they create a positive charge on the inside with a battalion of nasty positive charges outside threatening to rush in? They make the charge with something other than protons. Many chemical elements are positively charged in water: potassium, sodium, calcium, and magnesium, to name a few. By building up other types of positively charged chemicals inside the cell, microbes can use these other charge gradients to push against the proton onslaught.

Other acidophiles give up on the proton motive force altogether. Protons are just too out of control to be trusted for essential needs, so these acidophiles use positively charged sodium gradients instead. Instead of a proton motive force, they use a *sodium* motive force to make ATP and drive cellular functions. Instead of trying to bake a cake with napalm, they simply switch to batter. Of course, they still have to deal with the napalm (they're surrounded by it, after all), but at least they don't depend on taming it for sustenance.

These defensive measures are effective, but they're usually not enough on their own. Sometimes, the microbes just have to bail like sailors on a sinking ship.[3] They toss protons out of the cell like mad through transporter proteins. Each one of those transporters uses a lot of energy, which makes it expensive to live in a volcanic lake. Fortunately, the environment also supplies them with ample wealth: chemical gradients between the reduced sulfur spewing out of the volcano and oxygen diffusing in from the air are like candy to microbes. It gives them so much energy that they can afford to constantly bail out excess protons.

The true irony of the whole situation is that the high acidity is exacerbated by the microbes themselves. When they oxidize the volcanic sulfur, they make sulfuric acid, which is what got them into this whole predicament in the first place. They must look at humans causing climate change and shake their little heads, muttering, "amateurs." These little guys have been poisoning their own environment for billions of years.

Acid, of course, is only one of the threats to life at the bottom of a volcano. There's also the fact that every now and then the entire thing explodes. So what happens to the microbes when they get blown up? We know a lot less about that. For safety reasons, people tend not to do experiments inside active volcanoes. It's much safer to squirt acid into growing laboratory cultures than it is to rush into a volcano right before and after it erupts to see how the microbes are affected. However, a couple of people who have done this are Brian Hynek, at the University of Colorado, and Guillermo Alvarado, at the University of Costa Rica.

They have sampled Poás frequently enough over the years that they're developing a picture of how microbes are affected by eruptions. Sometimes, they find many different kinds of microbes in Poás Lake. Some of these don't even seem like they're acidophiles and might've been washed in from a rain event. My samples contained a similar variety of microbes—a mixture of those related to known acidophiles and others that look like stricken tourists who wandered into the bad part of town. Other times, Brian and Guillermo find just a single type of *Acidiphilum* has taken over completely.[4] At these times, a sterilizing event like an eruption may have created a new habitat that allowed this microbe to bloom into a nearly pure culture. Based on this research, it seems like volcanic lakes may oscillate between a teeming, diverse mass of microbes and a monoculture after each cleansing eruption. Personally, however, I don't plan on spending enough time down there to witness the changing of the seasons.

Living in Bleach

If microbial life can survive at pH levels near zero, what about the opposite end of the spectrum: Can they survive in an alkaline, high pH environment? Fortunately, Earth has subsurface features that can help us answer this question. We even call them volcanoes, though they're not, technically, volcanoes. These "mud volcanoes" (introduced in chapter 1) spew mud and fluids instead of lava or superheated water, and their eruptions are not driven by mantle melting like those of real volcanoes. In many mud volcanoes, life has the opposite problem from Poás. It must contend with very high pH values.

Environments that have the pH of cleaning products may seem rare—the products of very specialized geology. But Earth doesn't care what *we* think is normal. High pH environments are bog-standard, as long as you're talking about the subsurface, and especially if you're talking about early Earth. Today, high pH environments are common due to the pervasiveness of a process called serpentinization, which produces the chemicals that can be used by chemolithoautotrophs to produce cellular biomass. If chemolithoautotrophs are the "plants" of the underworld, serpentinization just might be their sun. Serpentinization may be even more important to life than the sun is, not because it supports as much total biomass as the sun—the plants definitely win in that regard. It is because, without serpentinization, we might never have even evolved plants at all.

Serpentinization happens when rocks from the upper mantle and lower oceanic crust are pushed into the upper 10 km or so of crust where they can react with water. This reaction produces minerals called lizardite and serpentinite—hence the name serpentinization. Serpentinization makes everything runny so the mud volcanoes flow and move chemicals throughout the subsurface. But perhaps most importantly for life, serpentinization makes massive quantities of hydrogen, a nearly universal food for microbes. Hydrogen can also react with carbon dioxide to produce methane and lots of other organic molecules that feed a variety of microbes. Serpentinization thus provides everything necessary for life, and all it needs is water, rock, and moderately high temperatures. This simplicity has led many to suggest that serpentinization might have fueled the first life on Earth.[5] It might also fuel life on other planets, asteroids, or moons.

Even though serpentinization is great for life, it comes with some headaches. Its chemical reactions drive the pH up to 10 or 12. In other words, serpentinization lays out a buffet of free food, but it's drenched in bleach.* Yet life seems to have found

---

* Most of the work bleach does for us, such as cleaning our clothes and sanitizing our bathtubs, actually comes from its strong oxidizing potential, not its pH.

a way to work through this. Dealing with bleach has some similarities to dealing with acid. In bleach, microbes don't need to
build dams to push protons out; they have to build dams to
keep them in.[6] They also constantly pump protons into the cell
at the expense of sodium ions that they pump out. As with
acidophiles, part of the solution for alkaliphiles (microbes that
prefer basic environments) is to just give up and deal with the
fact that the pH is going to be a bit high inside the cell, allowing
it to creep up to over 8.

As a result, many of these organisms can't really live at neutral pH anymore—they're permanently committed to the high
pH lifestyle. The real problem for these microbes, though, is
powering the cell. Recall that most life uses a proton motive
force, which it generates by pushing protons out of the cell to
create a gradient. When you live at pH 12, pumping a proton out
of the cell is a losing proposition. As the environment is basically devoid of protons, once you push a proton out, it just dis-
appears into the bleach. It will never fall back in and make your
ATP for you. There's simply no way for alkaliphiles to make the
type of acid gradient that drives power production for most of
the rest of life on Earth. One clever solution is to develop internal membranes, allowing alkaliphiles to pump protons into an
environment they can control. But cells that don't have internal
membranes just have to rely on the small amount of energy
from the charge gradient created when they push out the sodium ions. Even though serpentinization makes energy-rich
environments, the microbes who live there are nonetheless im-
poverished. The high pH won't let them obtain much energy
from all this good hydrogen that serpentinization produces. It's
like they've got a tapeworm.

The truth is, however, that we're still a bit in the dark as to
how microbes deal with high pH. Consider NPL-UPA2, the
name given to an uncultured phylum recently discovered at

ultrahigh pH serpentinizing subsurface springs.[7] Parts of this organism look like a hybrid of bacteria and archaea, making it a bit of a weirdo, genomically speaking. We currently have no idea how it deals with high pH. Since it's on a different branch of life than all known cultures, it's likely to have some wacky new ways of dealing with high pH that we haven't even thought of yet.

More Nasty Stuff

Acid, alkalinity, and an unstable environment are not the only stressors in volcanoes, whether those volcanoes are real or of the mud variety. Heat, salt, toxic metals, cold, desiccation, ultraviolet irradiation, radioactive irradiation, and pressure are also potent destroyers of living cells. Microbes use similar methods to tackle each of these challenges, with this general recipe: do what you can to keep the bad stuff out,* keep the good stuff in, and learn to tolerate a small amount of stressor that you can't completely get rid of. And for each of these other things, microbes show us that the full range of what life can handle is much, *much* greater than what we animals and plants can tolerate.

Most of what we know about how these guys tackle harsh habitats comes from throwing a single microbial culture into tough conditions and seeing what it does. But what about multiple stressors at once? These habitats usually have a mixture of ways to kill you. We need to do much more research to see how getting walloped with multiple different stressors at once affects these microbes. Also, in nature, these microbes are almost never on their own. Being in a rich community with lots of different types of microbes with different capabilities may help them to face high pressure, temperature, salt, metals, and acid all at once.

---

* This is notably not an option when dealing with heat, cold, or pressure. Cells can't keep those out.

There may be places to hide or ways the community members help each other that we haven't discovered yet. The specialized enzymes that help them deal with these harsh environments have historically proved to be useful in industry and biomedicine. I would bet good money that there are enzymes that will be just as useful to humanity as Taq polymerase—the enzyme that makes all of our DNA studies possible and came from microbes living in a hot spring. If we can learn more about the ways that intraterrestrials survive in volcanoes, we may discover ultraresilient enzymes or learn how to engineer other microbes to toughen up. In this way, life inside volcanoes might just enable our next big technological leap forward.

These extreme adaptations make it seem like intraterrestrials are playing by a different set of rules than the rest of us. Their family lines fall on cryptic deep branches on the tree of life that were invisible to us until revealed by recent advances in DNA sequencing. They hang out happily in places that would kill most organisms. And they survive on rocks and water alone. Does this mean that they break fundamental laws about what it means to be alive?

6

BREATHING ROCKS

UNFORTUNATELY, biology is not known for its absolute laws. There will never be an equation for who survives a volcanic explosion or why we evolved two arms and not four. When a species dominates an environment, it's tempting to assume it has the ideal traits to succeed in that environment. But maybe instead it spewed toxins, killing off better-suited microbes. Or maybe it was the one lucky duck that didn't get washed away in a rainstorm. It's nearly impossible to predict how an ecosystem gets from point A to point B from first principles alone because biology has an infinite number of ways to get the same result. It's not exactly a law-abiding citizen.

Here's an example of biology's lack of generalizability: when salmon stocks declined in the Pacific Northwest of the United States, scientists eventually determined that the drop in salmon was caused by a loss of kelp, which provides habitat for young fish. They then figured out that the loss of kelp was caused by an increase in sea urchins, which were eating the baby kelp polyps. In turn, the increase in sea urchins was caused by a drop in the sea otter population, since otters eat urchins. So if you want to increase the salmon population in the Pacific Northwest, you need to bring back the sea otters, which are cute, and everyone wants to have them around any-way. A win-win.

This is a great piece of scientific work, but it did not birth any universal laws. If we need to increase salmon stocks in Maine, for instance, we can't deploy sea otters to fix the problem because the East Coast of the US doesn't have the kelp-urchin-sea otter-salmon dynamic that the West Coast does.

But what if biology does, in fact, have fundamental laws, but we've been too focused on the details to see them? We know life is tied to energy and activity. Every living thing gets energy from somewhere and uses that energy to perform activities that change its environment. If we put energy and activity together, we get thermodynamics. Thermodynamics may sound like something that can safely be ignored if you're not a physicist, but if we're ever going to understand the true nature of life, we're going to have to extricate thermodynamics from its purely lifeless realm and give it the attention it deserves.

If thermodynamics is, in fact, the secret framework support-ing life, then intraterrestrials and extremophiles are the *only* beings on Earth that can show us where those boundaries lie. Intraterrestrials grab tiny slivers of energy that would be too small for us to use, and they get this energy from food that sounds more like it came from a chemistry lab than a restaurant. They do things we couldn't even imagine doing. Intraterrestrials are like the famous ballerina Misty Copeland. They have the control, nuance, and stamina to play Earth's chemical reactions like bril-liant works of art. We humans are more like babies bouncing our diapered butts to music. Sure, we're technically dancing, but it's a poor performance compared to the professionals.

It may seem like intraterrestrials don't have any limits at all (e.g., thriving in the deep, dark underworld and luxuriating in battery acid), but thermodynamics tell us there *must* be boundaries on all life; intraterrestrials are not exempt. However, every extreme place I've mentioned so far in this book has teemed with microbial life. So where on Earth, exactly, are the limits for these seemingly omnipotent organisms?

How about a place where the conditions are so intense that microbes can't gather up the energy necessary to survive? How about if we look for intraterrestrials in steaming jets of toxic volcanic gas?

Pica, Chile, 2022

Irruputuncu volcano in Chile is unlike Poás volcano in several respects. Instead of a lake, its crater contains dangerous gas jets. And instead of climbing down into the crater from a cushy visitor center to sample it, you must ascend the volcano—all 5,163 meters (16,939 feet) of it, mostly on foot. Nevertheless, Irruputuncu and Poás share the trait most important to our search for life's boundaries: an extremely inhospitable environment. This is why, one morning in March 2022, I found myself driving a truck near the border between Chile and Bolivia on the way to Irruputuncu.

My colleagues* and I awoke early in the small town of Pica and headed out in the dark, loaded with sampling gear, gas masks, food, water, and emergency oxygen bottles. Soon, we saw Irruputuncu looming in the distance, with a bright white plume steaming from its crater and no visible vegetation, in stark contrast to the lush tropical forest that surrounds Poás. The only building nearby was the Carabineros police station, where we stopped to show our permission to be there and were told to return when we came back down so they'd know we hadn't died on the volcano.

---

* This project was a multinational scientific collaboration with Chilean colleagues Gerdhard Jessen, Jenny Blamey, and Felipe Aguilera.

When we reached the base of the volcano, we dropped off two team members so they could measure gasses from a distance. But the only way to find out if there were any microbes living in the volcano was for six of us to hike up and into the crater itself. To save our lungs for the last bit of the ascent, we planned to drive our truck as far as possible up the rocky flank of the volcano. I intermittently gunned the engine over small boulders and nosed slowly into washed-out gullies until, with no warning or "road-closed-ahead" signs, I had to slam on the brakes to avoid careening into a vertical drop—Wile E. Coyote style. Flabbergasted, we climbed out of the trucks to see what we had almost fallen into. A steep-walled canyon sliced completely across the road, extending off into the distance on either side. This was a human-made trench, and if that wasn't enough to stop us, an insurmountable berm had been built on the other side.

Like many of the Andean volcanoes, Irruputuncu is also a national border, and this trench was intended to stop the movement of people between Chile and Bolivia. I realized that we had nearly fallen into a "border wall," not unlike the one that has alternatively been constructed and destroyed between the US and Mexico. We were forced to leave our truck behind and move more dangerously and slowly on foot, exposed to the elements. This made our work much more difficult, but I couldn't help but think of the people who face much higher stakes when crossing this border, hoping to improve their lives.

We began to trudge up the volcano. After every three or four steps, I would double over, gasping like a landed fish, then start again. We scrambled up giant piles of rubble, but as long as we avoided the steep scree, the climb didn't require any ropes or technical skills. Just lung power, and lots of it. After what felt like an eternity, we glimpsed white gas billowing out of a gash in the mountaintop. We had reached the crater of the volcano.

The crater was crescent shaped with a thick floor of garish yellow sulfur, a mineral that people had mined from the volcano decades earlier. The remains of their stone platforms were still standing in front of the crater and there was a wide opening in one of the crater walls, which miners had likely created. Through this, we were able to walk right into the gaping maw itself.

Inside, violent jets of gas shot up in evil-looking columns from black orifices. These jets reached the height of ten-story buildings, although the scale of the surrounding environment made it difficult to appreciate. Massive, house-sized boulders teetered at the crater's edges, threatening to clatter to the bottom, yet as I stared up at them, they looked no bigger than cows because I had no reference to compare them against. I was surrounded by nature in skewed proportions, and all of it looked like it wanted to kill me. The ground around the gas jets (which are called fumaroles) was bright orange like rotten fried eggs. Shouting over the deafening sound of the jets, my colleague Maarten de Moor told me that the different colors indicate how hot the gasses are. He poked a thick metal rod into one of them to measure the temperature. It was 408°C (765°F): hot enough to incinerate all the food particles stuck to your oven.

In addition to the heat and noise, the crater also bombarded us with thick gas. So thick, in fact, that we made a disturbing discovery: some of our heavy-duty gas masks, with black rubber straps and intimidating-looking metal canisters, were complete crap. The second I stepped into the crater, the gasses cut my lungs like a knife and singed my nose. I yelped in pain and ran out of the crater, ripping off the respirator and gulping clean air. I was lucky that I was able to exit the crater as quickly as I did because these volcanic gasses are more than just painful—they're toxic. In addition to containing high amounts of carbon dioxide that will asphyxiate you, they contain gaseous heavy metals and extreme acids like hydrofluoric acid, which, in sufficiently high concentrations, will dissolve your bones.

One of my colleagues had brought a much better mask with him because he needed to spend hours enveloped in the noxious gasses, painstakingly collecting them in a large glass bulb. When he took a break from his work and stepped out of the crater, he offered me his mask, which I gratefully accepted. As I watched him wipe his snot and spit out of the mask before handing it to me, I thought about how I had spent the past two years unwrapping a crisp new KN95 mask every time I went to the grocery store. It was oddly refreshing to have my COVID-19 fears take a backseat to the greater threat of bone-dissolving acid. I yanked the bulky respirator over my head and adjusted the thick rubber straps.

I stepped into the gushing crater like I was stepping onto the surface of the moon. I was cautious, lest my feet break through the thin crust into the scalding substances just below. Everything in my field of vision was blindingly yellow, shrouded in the hazy steam of the fumaroles, punctuated here and there by orange and black. The crater walls towered like a cathedral with spires of steam as buttresses. Moving through the thick cloud of noxious gases felt almost like scuba diving.

I looked around the crater, pondering what to sample. Extremophiles, as I explained in the previous chapter, can survive conditions that would kill any animal in seconds. However, even extremophiles can't handle 408°C. As far as we know, life hasn't yet found a way to overcome incineration. So I took a pass on the pitch-black holes-to-hell and made my way to the bright yellow sulfur crystals. They were beautiful up close, with delicate yellow spikes glinting like precious jewels in the sunlight. They were also soft and friable, so I could scrape them with a spoon. The hardest part was preventing the crystals on the spoon from getting knocked off by the jets of gas that came thundering out of the Earth like busted steam pipes. When I finally filled enough sterile plastic containers with this precious stuff, I exited the crater and passed the good mask back to my colleague so he could finish collecting his gasses. I then took a nice nap in the shadow of a rock just outside the reach of the fumes.

When everyone was done sampling, we began our descent, periodically checking to see if we were back in radio range, so we could notify the ground-based team that we were returning. We soon reached the top of a scree pile that we had studiously avoided on our ascent that morning. It looked like a gravel parking lot tilted at nearly a 70-degree angle. As we peered over its edge, my colleague Jacopo Pasotti turned to me, smiled, and said, "Wanna see the fastest way down a volcano?"

Before I could answer, he flung himself off the edge. I gasped as he landed in the nearly vertical rubble, planting his heels like ski edges. Then he did it again and again, bounding like Tigger until he was almost out of sight. I thought, "Well, I can do that too." I jumped down that volcano like it was an inflatable bouncy castle. At times, I went so fast that I lost control, sliding on my butt with the vague hope that I would avoid crashing into the boulders below.

Luckily, my samples were secure in my backpack, so everything I collected that day was ready for the long, difficult analysis when we got back to our laboratories. So did we, in fact, find a place too harsh for intraterrestrial life? For now, it seems that we did, since our DNA extraction attempts have yielded nothing. This may mean that the energy available for any living organisms at Irruputuncu is insufficient to "pay" for the effort it takes to survive in this environment. But that may not be the end of the story. Perhaps if I had taken a bigger sample, or if we changed DNA extraction methods, we would have detected living biomass. Nonetheless, thermodynamics tell us that a limit must exist. It's possible that we found it in Irruputuncu.

But poking around in volcanoes is not the only way to discover these restrictions. The power of thermodynamics lies in predicting where and when life is limited by energy and activity, even for intraterrestrials that operate way beyond the boundaries of human limitations.

How Do Thermodynamics Set the Boundaries for Life?

Thermo means "heat," which is a form of energy, and *dynamics* means "movement" or "change." Thermodynamics, therefore, is about moving energy around, which is the most essential function of life as we know it. When DNA unwinds so it can be copied for cell division, this is a chemical reaction. When a bacterial cell swims across a petri dish, this requires a whole cascade of chemical reactions. Almost everything a cell does, whether it's a microbe or a human cell, can be reduced to chemical reactions. To run these chemical reactions, every living cell needs energy. Each cell gets this energy from *other* chemical reactions.

Life's functions can be boiled down to this: life mixes up chemicals that release heat or energy (just like those little hand warmers that heat up when you shake them), then life grabs the energy from these chemical reactions and uses it to do wonderful, terrible, or mundane things. The rules for how energy moves between these chemical reactions are determined by thermodynamics.

One of the most important chemical reactions for powering life is respiration. The word "respiration" sparks strong emotions in those of us in the Animalia branch of life because we can't even go a few minutes without breathing. But what is respiration, really? Sucking air into our lungs is not really it. Lungs are kind of beside the point. Equating respiration with pulling air into our lungs is like equating dinner at a restaurant with placing an order with the waiter. You need to make the order to get the food, but the point of the meal is to eat it. In the same way, the point of respiration is not to suck in oxygen, but rather to run chemical reactions with the oxygen we've sucked in.

Thermodynamics determines whether the amount of energy we get from respiration is enough to power a cell to do the things it needs to do. Here's one way to think about this: respiration is like the first initial drop on a roller coaster ride. All roller coasters start with pulling cars up a big hill and releasing them at the top. The momentum produced from this drop powers the whole series of loops and twirls that come afterward, so the rest of the ride can never go higher than the initial hill, since that would require more energy than it started out with. In the same way, the energy that an organism obtains from respiration must be sufficient to drive everything else the cell does. The cell needs to move, defend itself, communicate with other cells, and reproduce. The bigger the initial drop, the more it can do.

In the late 1800s, a theoretical physicist named J. Willard Gibbs developed a concept for quantifying how energy moves around, and we can use it to figure out how much energy we receive from the initial drop of life's roller coaster. It's called Gibbs Free Energy, or ΔG (when you say ΔG out loud, you say "delta G"). The Δ symbol represents the difference between two chemical states—the top of the roller coaster and the bottom—so ΔG is the amount of energy given off when a chemical reaction occurs (when the roller coaster makes the big initial drop). When ΔG is negative, energy is available for life. When ΔG is positive, life has to spend energy to make the chemical reaction happen. A ΔG of zero means no chemical reaction can happen—in other words, death. ΔG is thus an extremely important value for examining how life works, as it tells us which reactions metaphorically put money in the bank, which reactions cost money, and which reactions can't happen unless something shifts their ΔG.

Since I'm suggesting that Gibbs Free Energy is the key to solving the mystery of life on Earth and possibly the whole universe, it's worth doing a bit of math to understand it. Even if, like me, you're not immediately drawn to math, stick with me because understanding ΔG is going to make you feel powerful.

ΔG can be expressed as the sum of two things. The first thing is $\Delta G°$ (pronounced "delta G naught"), which tells how good the chemicals are at reacting with each other. Think of those chemical hand warmers again. *Reactants* are the chemicals that go into the reaction when you shake up the warmers (in this case, iron powder and oxygen), and the *products* are whatever's left over after they react (iron oxide, or rust). $\Delta G°$ tells you whether the type of chemicals you have are likely to react. The chemicals in hand warmers (iron powder and oxygen) are inherently good at reacting with each other so they have a negative $\Delta G°$. You can't make hand warmers out of beach sand because it's not naturally reactive with oxygen.

The second thing you have to add to $\Delta G°$ to sum up to the total energy of ΔG is a number that tells you how much reactant you have left to use up. A fresh bag of unreacted chemicals has more potential energy than one that's almost used up. The variable Q (the reaction quotient) tells you how much of the reactant has already turned into product. This gives you the full Gibbs Free Energy equation:*

$$\Delta G = \Delta G° + RT \ln Q$$

---

* The other variables in this equation, R and T, stand for the ideal gas constant and the temperature in Kelvin, while "ln" is the "natural logarithm" symbol. All these things are easy to put into an equation, so we don't really need to think about them for our purposes here.

It's pretty, right? Even if you don't think it's pretty, maybe you'll agree that it's practical. This compact little equation is all you need to determine how much energy is available for life. By offering us a real, useful rule for life, this equation can point us to where new possibilities exist. It just so happens that this equation predicts that there is energy available in the fumaroles of Irruputuncu volcano, as well as the moons of Jupiter and many other extraterrestrial bodies. If this energy is enough to fuel the activities necessary to survive, then all we have to do is look for the life that's using it and, voilà, we'll discover new types of life on Earth or extraterrestrially.

How Does Life "Catch" Energy from Thermodynamics?

A good way to think about how life harnesses energy from a negative ΔG is to think of each chemical reaction as starting with a ball sitting at the top of a hill. When the chemicals react, the metaphorical ball rolls down the hill and emits energy. Some of that energy is lost to friction, but you could also set up a series of paddles that "harness" the energy from the rolling ball. If you connect those paddles to a turbine that charges a battery, you can use the battery to power whatever you want.

In the biological makeup of most living things, the paddles and turbines that harness energy are enzymes, small molecules, and chemical gradients like the proton motive force I discussed in the previous chapter. And they are powered by chemical reactions, rather than balls rolling down hills. But energy is energy, so energy from the ΔG of a ball rolling down a hill, calculated from the weight of the ball, the height of the hill, and the acceleration of gravity, is mathematically equivalent to the ΔG of a chemical reaction, calculated with the equation above. You can even use ΔG to convert the energy of respiration to something as strange as pineapples falling off a roof. To get enough energy to "throw my hands up in the air sometimes," as the singer Taio Cruz does, I'd have to capture the energy of a pineapple falling off the roof of a two-story house. To power a long and satisfying human life, full of love and cherished moments, I'd have to chuck four billion pineapples off the roof. It may seem dreary to reduce all the wonderful things in life to units of pineapples falling off houses, but it pulls back the curtain on the thermodynamics quietly driving everything in the background.

In human respiration, the energy gradient that gives us our ΔG is the chemical reaction between oxygen and sugar that "reduces" the oxygen and "oxidizes" the sugar. This is called a redox reaction because the first chemical gets **RE**duced as it accepts excess electrons from the second chemical that gets **OX**idized. For us, puppies, fungus, and most everything else in our daily lives, the redox reaction between oxygen and organic matter is our only option for getting ΔG. The reaction is highly energetic, so our $\Delta G°$ is extremely negative, and we can use it to do all sorts of complicated business, from tying our shoes to winning the World Cup.

Therefore, it may sometimes seem like life depends on oxygen, but it doesn't. It may also sound like redox has to do with oxygen, but it doesn't. Gibbs Free Energy sets the rules, and the rules say that respiration of oxygen is not the only chemical reaction with a negative $\Delta G°$.

In the gritty underworld, oxygen is hard to come by. Not that subsurface microbes necessarily want it anyway. Oxygen is toxic. Its extreme reactivity causes all sorts of problems for the other chemical reactions that a living cell needs to do. Those of us who breathe oxygen make an army of enzymes that scurry around our bodies detoxifying the chemicals produced by oxygen when it runs amok. We work hard to keep our respiration from killing us.

Gibbs Free Energy shows us that oxygen is a minuscule slice of the total amount of chemicals whose energy yields can be harnessed to support life—consequently, life can breathe many weird things. In fact, "respiration" has nothing to do with breathing oxygen, specifically. I'm sometimes forced to reeducate the students in my undergraduate lectures by calling out, "Does respiration mean oxygen?" repeatedly until all the "Yes's" change to "No's." The truth is that almost any chemical redox reaction, even those that do not involve oxygen, is used by some type of life to get energy. All of these redox reactions constitute respiration for the types of life that use it, even though we humans would never be able to breathe the chemicals ourselves. Intraterrestrials respire other, less powerful chemicals that we don't usually think about, exploring spaces within the thermodynamic constraints of life that the rest of us can't. Here, energy margins are often slim, skating at the edge of what is thermodynamically possible.

Expanding the Boundaries of "Breathing"

So if they're not breathing oxygen, what are they breathing? Some of the most popular things to respire are nitrogen and sulfur compounds of different oxidation states, meaning they each yield slightly different energies than each other. Commonly, these include nitrate (fertilizer), nitrite (toxic to fish), nitrous oxide (laughing gas), and sulfate (abundant in seawater), as well as sulfite and thiosulfate (such hot commodities that they rarely build up in natural systems because they're so valuable). Carbon alone makes such a bewildering array of breathable compounds that we definitely haven't discovered them all yet.[1]

The nonexhaustive list includes fumarate (fruity-flavored, often added as a preservative to food because it's harmful to most microbes), dimethylsulfoxide (sunscreen made by algae), trimethylamine (a fishy odor), and chlorinated carbon molecules (used to clean fighter jets).[2]

Perhaps the most surprising carbon chemical that can be respired is carbon dioxide. Carbon dioxide is our trash. We can't do anything with it besides get it out of our lungs as quickly as possible. But to microbes called methanogens and acetogens,* carbon dioxide is as good as pure, life-giving oxygen is to us.

Thermodynamics show us that, while the idea of breathing carbon dioxide may feel like combing our hair backward, it's A-OK for supporting life. In fact, counterintuitively, most of the periodic table can be "breathed" by intraterrestrials. Radioactive uranium doesn't really have a "life support" ring to it, yet that's just what it is for some intraterrestrials. Arsenic is rat poison and a favored murder weapon for mystery writers and real murderers alike, but intraterrestrials respire it too, occasionally cleaning up our toxic pollution. Gold is not very chemically reactive; that's why humans have historically used it in currency and jewelry. Yet there are microbes that take advantage of the Gibbs Free Energy available in reducing dissolved gold into solid gold. Many solid metals, it turns out, are great things to breathe, including iron, manganese, selenium, vanadium, and chromium. So technically, intraterrestrials can "breath rocks."

---

* Plants, of course, also use carbon dioxide, but they don't get energy from it. Instead, they put energy into making sugar and oxygen by capturing the sun's photons.

There are at least twenty elements[3] on the periodic table that some microbe somewhere has evolved the ability to respire.* In fact, these alternative modes of respiration are widespread and common, even if we mostly don't notice them. Breathing rocks may have been more prevalent in Earth's history than breathing this toxic oxygen stuff. If the hard-rock-breathing intraterrestrials were writing this book instead of me, they might be shocked to discover that we breathe a *gas*. Maybe *we're* the ones who are weird. Oxygen only showed up halfway through Earth's history. Microbes have been breathing metals the whole time. This is great news for us, because some of these respirations can remediate our industrial waste and toxic pollution, whenever we can coax microbes into doing it for us.

Ready to see how far we can push the boundaries of thermodynamics? What if some intraterrestrials breathe things that are not even chemicals at all? Chemicals are made of atoms, and atoms are made of protons, neutrons, and electrons. When life "breathes" a chemical, the electrons do all the work, jumping between chemicals to make redox reactions happen. It turns out that some types of life are apparently fed up with the bulky chemical folderol. This life, of which intraterrestrials are major players, strips energy down to its bare essentials: pure electrons.

There are many ways that microbes use pure electrons. Some make special structures called nanowires that work like tiny electrical wires to create a current of moving electrons.[4] Others use metal-containing proteins to make little electron highways. Some of these microbes pack so many of these metallic proteins

---

* These include hydrogen, vanadium, antimony, chromium, manganese, iron, cobalt, nickel, copper, palladium, oxygen, nitrogen, carbon, fluorine, chlorine, sulfur, bromine, selenium, arsenic, iodine, gold, platinum, and others.

together that their cell wall is more like a metal sheet than a normal biological membrane. Here, electrons fly out in every direction so the whole cell can access the energy nearly instantaneously. For other microbes, the exact mechanism is not clearly worked out, and we just call it direct interspecies electron transfer (DIET). When I listen to the scientists who study these DIET microbes talk about their work, they sound more like electrical engineers than biologists, throwing around words like cyclic voltammetry, cathodes, and anodes. Life has long known that even though electrons are only a ten billionth of the mass of each atom, they are the seat of power. When distilled to the thermodynamic requirements for life, it makes sense to cut out the chemical middleman and go straight for the electrons. The advantage is that you don't have to wait for chemicals to diffuse into the cell. You could cut the end off an extension cord and use it like a snorkel instead.

Speed Matters Too

Not all energy-yielding chemical reactions can fuel life. In some cases, the reactions are simply too fast. The speed of a reaction is called its kinetics, which is distinct from but just as important as thermodynamics. It's almost as if some chemicals have an innate desire for each other and will react too quickly for life to get a chance to catch the energy. In other cases, however, the reaction is sluggish. This is life's time to shine. It's as if the reaction is willing to "pay" microbes to help the chemicals react, rewarding them with energy in return.

If you fly to New York to see a Broadway show, for example, technically you can walk the ten miles from the airport to the theater. But chances are you'd much prefer to take a cab or the subway to speed up the process. Life speeds up chemical reactions that "want" to go but are, for whatever reason, prevented from doing so quickly. You can think of all life on Earth as a bunch of taxi drivers that transport chemicals to where they want to be. And, like taxi drivers, all life on Earth gets paid for its labor. This "payment" comes in the form of those proton gradients that I described in the previous chapter that create ATP and internal energy storage that cells can use— for instance, to fight acid and bleach. But if the chemical reaction happens fast enough on its own, life is out of a job. This thermodynamic view almost flips our whole conception of life upside down—instead of life harnessing energy from chemical reactions, maybe life evolved because slow chemical reactions needed catalysts to help the reaction happen. Maybe the chemicals themselves are our overlords.

Here is another example to deepen the analogy above: consider the reaction between sulfide and oxygen. The sulfur atom in sulfide is packed with as many electrons as it can handle, so it's looking for a place to offload them. Oxygen, on the other hand, is one of the most electron-hungry molecules in the universe, so it will gladly accept electrons from sulfide. Under most conditions, the two chemicals react far too quickly for life to get involved. But there are many places in nature where sulfide and oxygen are separated physically and can't react. For instance, in smelly marine mudflats near the ocean, sulfide is buried down in the mud, out of reach of the oxygen in the air. In such places, you'll sometimes see thin, white, wispy strings poking out of the mud. These are filaments of bacteria that partially oxidize sulfide, turning it into solid elemental sulfur, which they then carry to oxygen as they glide vertically in their sheaths, like bees transporting pollen between plants. When the sulfur and the oxygen finally meet, they are able to react, and the sulfide-oxidizing bacteria keep the energy that results from the reaction. Just like well-paid taxi drivers.

The Rules for Life

Given the incredibly mind-bending nature of intraterrestrial life, why bother to learn these underlying thermodynamic rules? Can't we just keep discovering new capabilities for life, bit by bit, without worrying about the unifying principles that drive them? Not if we hope to advance our understanding beyond what we know today. Understanding the thermodynamic rules for life gives us the power to predict where else we may find it—whether that's a strange place on Earth or a different planet entirely.

So here they are, the thermodynamic options as they apply to life:

1. Processes that yield energy (ΔG is negative) but are slowed or blocked. These are great at supporting life. In environments where this is the case, we should expect to find life.
2. Processes that yield energy (ΔG is negative) but happen so quickly that life doesn't have a chance to participate, or they produce an amount of energy too small to be useful. These are not useful for life; they are abiotic reactions. Fire is an example of these types of reactions.*
3. Processes that require an input of energy (ΔG is positive). These cannot yield energy for life. If a cell needs to run this reaction, it will have to spend the energy that it captured from a different reaction or manipulate Q to change the sign of ΔG.
4. Processes that neither yield energy nor require it (ΔG is zero). This is death.

---

* Everett Shock and Eric Boyd put this more colorfully, "Things that burst into flame not good to eat" (Shock, E. L., & Boyd, E. S. Principles of geobiochemistry. *Elements* 11, 395–401 [2015]).

That's it. The thermodynamic boundaries of life are not complicated, but they're powerful. They show us that, even though it may seem like we've already got an exhaustive list of processes that support life, there are far more possibilities yet to be discovered. Every time a scientist has said, "Hey, this is a chemical reaction that could support life, I wonder if it does . . . ," it eventually leads to a great discovery. Discoveries of this sort pop up every few years, like little gifts for geobiologists. And some have fun names. Anammox, for instance, was predicted in the 1970s, discovered in the 1990s, and sounds like a Marvel villain. Anammox is the oxidation of ammonium, a waste product from other microbes, coupled to the reduction of nitrite, a nitrogen compound of intermediate oxidation state, plentiful in Dutch ditches.[5] The microbes that breathe with anammox produce hydrazine while they're doing it. Hydrazine is rocket fuel, originally used to power fighter jets in World War II, and now used for more peaceful operations such as maneuvering spacecraft on Mars.

We have every reason to believe we'll discover more types of respirations in the years ahead.[6]* As Dimitri Mendeleyev, the creator of the periodic table, used the quantized patterns of subatomic particles to predict the existence of elements that were only found years later, we can use what is known about redox chemistry to predict new metabolic redox pairings that might support life. And one nice thing about these thermodynamic rules is that they are not Earth-centric. When they show us an open door for life on Earth, that same possibility extends to life outside of Earth.

Breathing weird new stuff is impressive, but it's not the only thing intraterrestrials do with their freewheeling approach to thermodynamics. Hang on tight. Life is about to get much, much weirder.

---

* See Lu et al., 2021, for a nice overview.

7

LIVING ON THE EDGE

TO UNDERSTAND this new level of weirdness, I need to introduce a concept for which we have no analogy in our daily lives. I need to talk about a "thermodynamic landscape." In the previous chapter I pointed out that it is possible to breathe something other than oxygen—even rocks themselves, simply because breathing rocks is thermodynamically and kinetically advantageous. Breathing rocks may be unexpected, but it's not a hard concept to understand. Unfortunately, we have no human parallel to the concept of a thermodynamic landscape or how it can shift around a living being. So you and I are going to have to go out on a limb together to understand it.

The thermodynamic landscape is determined by the array of chemical reactions available to support life in a given environment. Because they often live at the edge of thermodynamic possibilities, intraterrestrials experience changes to their thermodynamic landscape in ways that can't happen to us. We, and most of the life we directly interact with, either breathe oxygen or photosynthesize. Since these are high-energy processes, nothing ever really changes about the thermodynamics we experience. Running out of food will kill us, as will lack of oxygen, but our thermodynamic landscape doesn't shift. A grain of rice will still give us the same amount of energy whether or not we were starving to begin with.

If I wrack my brain for an analogy for what it must be like to experience changes to one's thermodynamic landscape, the only thing I can think of is travel. Have you ever traveled to a new place and felt like everything you assumed was true just disappeared? This must be what it's like for an intraterrestrial when their thermodynamic landscape shifts: suddenly their respiration no longer yields energy, even if their reactants are still present.

I had one of these jumps in perspective when I was in Seoul, South Korea. During dinner with my colleagues, a flurry of waiters brought out small dishes and whisked them away after we emptied them. The dishes were thrilling and unfamiliar to me. There was a vibrant pink soup, artfully arranged vegetables, and a single butterflied shrimp on a lattice made of a chewy substance. As I happily gnawed on the lattice, a waiter rushed over to me, his cheeks flushed with embarrassment for me, and told me it was plastic. The next night, I ended up in another banquet, chatting happily with my colleagues, while waiters brought in plate after plate. I looked up to see tender chopped beef on a bed of what can only be described as sticks. I thought, "Aha. Yesterday I was a noob. Today, I know what's food and what's not food. This is a pile of sticks. Sticks are not food." And, like the savvy international traveler that I am, I chopsticked the beef, and left the pretty stick display on the plate. The waiter looked confused as he noticed my neglected pile of sticks, so I dutifully ate them. They were delicious.

For the intraterrestrials, getting buried to a new depth layer, being thrust upward in a mud volcano, or facing countless other environmental changes shifts their thermodynamic landscape in ways that are probably as jarring as eating a banquet in a new country. In this chapter, I will describe what it's like to work in Svalbard, Norway, where the intraterrestrials have a unique opportunity to alter their thermodynamic landscape. Then I'll show you what they do with these strange thermodynamic situations: they wage thermodynamic warfare, acquire energy by alternating breathing forward and in reverse,* and push life to the absolute limits of energy deprivation.

Svalbard, Norway, 79°N, 2016

Even though it was the middle of the night, I squinted in the bright sunlight as I turned to my PhD student, Joy Buongiorno, and said, "You doing OK? Should we go to sleep, or start processing the next core?" The sun does not set in July at 79°N, which is north of most of Greenland, Canada, and Russia. Instead, over the course of each twenty-four-hour period, the sun crawls along an elliptical track in the sky. If Joy and I were scientists on TV, we would just put our sediment cores into a giant machine that spits out data. But real science doesn't work like that. We were spending hours painstakingly slicing the cores into tiny sections and preserving them for the measurements we would make in our home laboratory. With our six hours of jetlag, a brilliant midnight sun, and the knowledge that the sooner we processed the cores, the more pristine the data would be, Joy and I were not inclined to go to bed. She replied, "I'm good if you're good. Let's start the next core."

We were working at a makeshift table in the tiny research town of Ny-Ålesund on the shore of Kongsfjorden, part of the Svalbard archipelago in the Arctic Ocean. Massive blue-white glaciers and towering mountains surrounded us on all sides.

---

* This might sound dangerously close to a perpetual motion machine, but don't worry, no laws of physics were harmed in the writing of this book.

The mountain peaks were strangely shaped, with pyramidal caps left over from the differential weathering of their rock layers. These crown-like peaks give the fjord its royal name, which translates to "King's Bay." The occasional Arctic seal and fox came by to check us out, while short fuzzy reindeer meandered by and didn't seem to notice anything that wasn't moss or grass. Giant icebergs, of a similar color to the Blue Razzberry drinks my kids enjoy, slowly floated past us, and northern fulmars—gull-like seabirds with fat bodies shaped like passenger jets—skimmed the water's surface like artists painting with air. There are no trees or large bushes on the archipelago, so the breeze was silent except for the gentle lapping of water. The only other sounds were the rhythmic ripping open of plastic Whirl-Pak bags, the soft "*chnk*" of our crimper as we sealed dissolved gasses into glass vials, and the sucking of snot back into our noses after we had wiped them on our jacket sleeves.

Joy and I were being hosted by the French-German station, AWIPEV, and the meals for everyone in town were being cooked by Kings Bay employees, who organize the logistical needs of the scientists. Svalbard is an archipelago that is the northernmost human outpost in the world, 2,000 km (1,200 mi) north of Norway and 800 km (500 mi) south of the North Pole. It is not technically in any country, which makes it difficult to fill out my university's travel forms. In practice, being in Svalbard feels like being in either Russia or Norway. This is because the Treaty of Versailles, signed after World War I, granted Norway control over the territory, with Russia also having rights. There are two main towns, Longyearbyen and Barentsburg, which each have less than a thousand people, a couple of stores and restaurants, no stoplights, and a lot of polar bears. The total population of Svalbard is about three thousand, roughly the same as the population of the polar bears.

The town of Ny-Ålesund is much smaller, with a permanent population of only a few tens of people. Besides science, the other goal of the people in Ny-Ålesund is not getting eaten by a polar bear, which takes some effort. The "town" consists of a few buildings, a power plant, a handful of vehicles, and a mass of yapping sled dogs. If you see a polar bear, the protocol is simple: run into the closest building—all buildings are required to be unlocked at all times for this purpose—and call the watchman to alert everyone else. When you're farther than a quick sprint to the nearest building, the situation is more complicated. At least one member of the group must be trained in how to use a high-powered rifle and flare gun. When you reach the edge of town, marked by a heavy dirt berm, you are required to stop and load four .308 caliber rounds into the magazine of your rifle. When you arrive at your research site, one person in the group is prohibited from doing any scientific work, as their job is to scan the horizon for bears. When I'm on polar bear duty, I always feel like a dilettante, standing there while everyone works around me. Nevertheless, it's a small price to pay to avoid being mauled.

Speaking of aggressive animals, the Arctic tern puts up a fight too. This is a small bird that migrates the full length of the globe every year, converting highly efficient aerodynamics into a ticket to perpetual summer. Svalbard is mostly uninhabited, but Arctic terns don't want to live in the uninhabited places. They want to live right in the middle of our research town and dive bomb us constantly, pecking our heads. Before I had experienced the terns myself, my colleague and husband Drew Steen returned from a research trip with amazing close-up photos of them. When I side-eyed him for getting close to this nearly endangered species, he protested, "I was using the camera in self-defense!"

Thankfully, Joy and I weren't being attacked by terns or polar bears when we pulled that all-nighter in the blazing sunlight. Instead, we were treated to a stranger sight—one that would change the course of our research. As we methodically cataloged and preserved every gram of sediments from these cores, we watched a mysterious tide creep across the fjord. The water, which was reddish brown in the sunlight and dark red in the shadow of clouds, looked like it was slowly taking over the whole fjord. Since a biblical plague was unlikely, we just kept working. We later learned that this red water was the out-flow from the Bayelva River, which picks up iron-rich deposits from summer glacial melt. We were looking at the very stuff that made up the sediments we were working on, material that tweaks the thermodynamic landscape of intraterrestrials in unexpected ways and upends some of our more basic assumptions about life. It seems, for instance, that life should be slow in the Arctic since it's cold and often dark and frozen. But intraterrestrials don't care about sunlight or cold. They care about ΔG. In ul-tralow energy environments, this bright red iron brings ener-getic firepower that fuels the intraterrestrials' arsenal when it's time to fight.

Thermodynamic Warfare

No single microbe living in Earth's crust can breathe all the dif-ferent chemicals that are available for respiration; they divide the chemicals among them. This is not out of courtesy. It's because there's a cost to maintaining a gene in one's genome, so microbes have to specialize. This division of labor presents in-teresting opportunities for competition. I have been writing about the Gibbs Free Energy equation as if the only thing you can do with it is figure out how much energy you have to work with. But the intraterrestrials are not passive wallflowers. They put ΔG to work for them, manipulating it for their own advantage. In the surface world, we know a lot about how microbes fight each other. They secrete deadly antibiotics, hoard nutrients, and grow ultrafast to get ahead. Intraterrestrials do these things, too, but they also fight each other with thermodynamics. If one microbe's ΔG is better than another one's, then the first microbe can *asphyxiate* the second.

This sounds pretty gruesome. But these thermodynamic fights happen all the time in the muck underneath our oceans, lakes, and estuaries. The biggest badasses in these thermodynamic fights are the microbes that breathe sulfate—the "Arctic terns" of the subsurface, if you will. Sulfate is a colorless, odorless chemical that floats around the worlds' oceans like a couch potato. In our oxygen-rich surface world, sulfate is quite dull. We drink it in tap water and barely notice it. But intraterrestrials bring sulfate to life. When they breathe it, they turn it into sulfide. Sulfide, as opposed to sulfate, is impossible to overlook. Sulfide reacts with iron to make black sludge or shiny fool's gold and kills small animals. It smells like rotten eggs, so it makes coastal mudflats reek. It sours oil wells and eats into the hulls of ships, wreaking havoc on human enterprise. But producing sulfide is only one of the reasons why sulfate reducers are the queens of the underworld. The other reason is that they get truckloads of energy through their superior $\Delta G°$.

To understand why $\Delta G°$ matters, let's go back to that equation for a moment: $\Delta G = \Delta G° + RT \ln Q$. As I discussed in the previous chapter, $\Delta G°$ is not the whole equation, but it's a big part of it. Having the best $\Delta G°$ means that sulfate reducers get more energy from breathing then the other microbes do when they all have equal access to their substrates. What should an organism do when it's got the best $\Delta G°$ around? Should it grow really big? Probably not, since getting too big means that sulfate can't quickly diffuse into the cell during respiration. Should it release an arsenal of antibiotics and annihilate the competition? This is also a bad idea. Microbes depend on their neighbors for things like food and vitamins, so killing them all would backfire. They only want to kill their direct competitors.

Who are their direct competitors? Any microbe that eats the same food as they do. Remember that to breathe, the electrons that reduce the oxidant (for us, the oxidant is oxygen, but for sulfate reducers the oxidant is sulfate) come from food. We humans get these electrons from sugars and other organic molecules. The food for sulfate reducers is usually very simple, like acetate, formate, or what may be the simplest food on Earth: H_2, a dissolved gas called hydrogen—the same one made by serpentinization, but its global production is much broader than just this one geological process. Hydrogen is a waste product of many organisms in the subsurface, so it's constantly being produced. And it has a special property that most other food doesn't—it passes freely in and out of living cells because it's small and doesn't have an electrical charge. Microbes can therefore snap it up like samples at the grocery store.

Since sulfate reducers are the big dogs, they could just take away the hydrogen and starve their competitors. But because hydrogen is small and uncharged, nobody can physically hold on to it. It floats in and out of cells like a ghost through walls. For microbes, hydrogen is unhoardable. Instead, sulfate reducers do something a little craftier, like a thermodynamic life hack. Since everyone in this environment is living close to the absolute minimum of energy that can support life, everyone needs about the same amount of total energy (ΔG).

However, because sulfate reducers start with a better $\Delta G°$ than everyone else, they can play with the rest of that equation, $RT \ln Q$. Nobody, not even sulfate reducers or humans, can do anything to change R, T, and ln. R is the universal gas constant, T is temperature, and ln is the natural log mathematical transformation that's just a button you press on a graphing calculator. What sulfate reducers *can* do is manipulate Q.

To calculate Q, we multiply the activities* of all the products together and divide them by the activities of all the reactants, with each one raised to the power of the number of times they appear in the chemical reaction. Sulfate reducers can't change R, T, or ln, but they *can* change the activities of chemicals in their environment. They do this by consuming them at a certain rate to maintain them where they want them. Since sulfate reducers have the superior $\Delta G°$ they *can reach the same ΔG at a lower ratio of reactants to products, Q.* This means that when sulfate reducers are running at peak efficiency, they can bring hydrogen levels so low that they make their competitor's ΔG positive. And remember, you only get energy from a negative ΔG.

In other words, sulfate reducers snuff out their competitors' ability to breathe by decreasing the concentration of the available hydrogen. They don't use it all up—they still need hydrogen for themselves, and it continues to be produced by other microbes.[†] They just pull it down to such a tiny concentration that the math doesn't work out for a microbe with a lower $\Delta G°$ starting point. Even though the sulfate reducers' competitors are surrounded by life-giving food, they can't eat it because the ratio of reactants to products, Q, is so low that their respiration doesn't yield energy anymore. Like a

---

*A chemical's "activity" is similar to its concentration, but it takes into account how all the other chemicals in the solution affect the chemical's availability for reactions. For example, sometimes salt will "hold on" to a chemical, so even if a salty solution and a freshwater solution have the same concentration of that chemical, the chemical will be less reactive in the salty solution.

† These are mostly the fermenters. More on that later.

shipwrecked sailor dying of dehydration while surrounded by water, these microbes expire with their food right in front of them. Sulfate reducers win because they take the whole system to the bitter edge of their own thermodynamic capabilities, which pushes everyone else off the cliff.

But sulfate reducers don't win forever. The mud and sediments where all these thermodynamic fights take place are formed by the slow raining down of muck from rivers and oceans to the seafloor, under which the sulfate reducers live. That means that sulfate reducers are slowly buried farther from the source of sulfate that sustains them. As a result, sulfate reducers only have a few hundreds to thousands of years to celebrate their victory. Once the sulfate is gone, the sulfate reducers can't breathe anymore, so the hydrogen builds up and their competitors can seize the opportunity.

The main competitors are methanogens, which are microbes that make methane. When the sulfate is gone, these guys are free to spew methane with wild abandon, as long as there's enough hydrogen and other simple foods like acetate, formate, and methylated compounds left over to feed them. Though it's tempting to cheer for the underdog, keep in mind that methane is a dangerous greenhouse gas, so methanogenesis is not necessarily a good thing. Fortunately, one nice side effect of this thermodynamic competition is that sulfate reducers prevent methane from being produced in shallow sediments where it could more easily leak out into the overlying water and possibly into the atmosphere.

Given sulfate reducers' usefulness in helping to keep methane far from Earth's atmosphere, one might wonder whether there are places on Earth where these tiny warriors get a boost. This is where the mysterious red water that Joy and I watched swirling around the fjord in Svalbard comes in. When the bright red glacially scoured iron settles on the muddy bottom of the fjord, it regenerates sulfate by chemically oxidizing sulfide almost as quickly as it is produced. Under these conditions—as my wonderful postdoctoral advisor, Bo Barker Jørgensen of Aarhus University, Denmark, has spent years demonstrating—instead of breathing up all the sulfate until it's gone, sulfate reducers in Svalbard's fjords get their sulfate recycled for free. It's like they have access to a perpetually stocked fridge. Bo and his many students and postdocs over the years have shown that even though the Arctic is cold and dark, its subsurface is alive and well, supercharged by the energy released from chemical reactions between iron and sulfur.[1,2] As a result, methane in these environments is pushed even deeper into sediments than it would be normally, keeping it farther away from our atmosphere and farther from potentially exacerbating our human-made climate change.

Reversible Breathing

Arctic fjord sediments are not supercharged for everyone. Things aren't so rosy for the microbes who get pushed over the thermodynamic cliff by the winners. But remember, intraterrestrials aren't like you and me. They can be more creative with their physics. Some of them can get energy from respiring in the reverse direction.

Let me give this point its own paragraph: *There are places in Earth's crust where a single chemical reaction that normally yields respiratory energy in one direction reverses and starts giving off energy in the reverse direction.* If life can take advantage of these reversals, it will be able to get energy from "breathing in reverse." There are no parallels for this in the surface world.

Sulfate reducers, as I described above, can pull hydrogen concentrations so low that they prevent methanogens from getting ΔG energy. But sometimes, when the conditions are right, they pull hydrogen so low that the ΔG for methanogenesis can actually shift from negative to positive and allow methanogens to gain energy by breathing in reverse. In this situation, methanogens take their normal waste product, methane, and use it to produce their normal reactants, hydrogen and carbon dioxide. The chemicals that were once their nutrients become their wastes and vice versa. This cannot happen in the surface world. It's as if you've fallen off a boat and are sinking in water, but instead of swimming back up to the surface to save yourself, you dive deeper toward certain death only to discover that once you get deep enough, you can breathe the water. It's like traveling down a number line: the numbers will become smaller and smaller until they cross zero and start going up in absolute value again.

If the intraterrestrials can get energy from breathing in reverse when their products build up, could we humans do it too? If we did this neat trick of pushing Q far enough to flip the sign of ΔG, we would actually *get* energy from breathing in carbon dioxide, spitting out a lump of sugar, and gaining enough energy in the process to sing an aria. It would be great! Sadly, we can't do this. If I took a balloon the size of the entire Earth and filled it completely with carbon dioxide molecules, then added six measly molecules of oxygen and a single tiny molecule of glucose, thereby wrenching this massively energetic reaction very hard into the reverse direction by skewing Q, I *still* couldn't get the reaction to yield energy in the reverse direction. In fact, even with this unachievable imaginary experiment, I would still get quite a bit of energy from running the reaction forward and adding to the already-enormous pile of carbon dioxide. The $\Delta G°$ for aerobic respiration is so negative that, at physiologically relevant temperatures and pressures, there's no amount of changes that can be made to the amounts of reactants and products that will allow energy to be gained from running the reaction in reverse. So we'll never live the dream of fueling a marathon while coughing up energy bars made inside our bodies. But the intraterrestrials can do this, at least for their version of marathons and energy bars.

We know that such thermodynamic reversals are possible for methanogens, but an important question remains: Are these microbes equipped to take advantage of them and grow equally well in either direction? Proving that microbes acquire energy coming and going is difficult since it's hard to grow these sluggish microbes in the artificial environment of a laboratory. One growing culture in the 1990s was able to reverse its breathing,[3] but it was subsequently lost. More recently, methanogens have been observed removing the methane they just produced.[4] And when I dredge up ocean muck for my own research, I see the same methanogens in areas where methane is being produced and areas where it is being removed.[5] For now, then, it seems that life has indeed found a way to take advantage of these energetic reversals. And why not? If there's one thing that we know about intraterrestrials, it's that if energy can be gained from speeding up a slow reaction, some microbe somewhere has figured out how to grab it.

These reversals almost seem like a perpetual motion machine, which is a thermodynamic no-no. However, because this reversibility is driven by changes in their external environment that alter ΔG for them, the microbes cannot simply produce energy forever. When concentrations of chemicals shift in their environment, it's like an external kick to the metaphorical fly wheel that keeps the whole process cycling back and forth over thousands of years.

My belief that microbes get energy through energetic reversals is not the consensus of the scientific community. When forced to give an opinion about whether respiration can reverse, many scientists' first inclination is that this reversibility theory must be wrong. Life simply cannot reverse directions and start breathing its own waste products . . . because . . . it just seems unreasonable. In addition to being unreasonable, it's also hard to experiment on, since the shifts in respiration can be subtle, so more work needs to be done to be sure. But according to thermodynamics, at least, this extreme option must remain on the table. When it comes to the intraterrestrials, we have to throw common sense out the window and replace it with thermodynamics. Of course, sloshing a chemical reaction back and forth is not the only solution to low-energy living. The intraterrestrials have other tricks up their sleeves.

Low-Energy Life Powers Up

Life at low ΔG doesn't have to be a drag. There are ways to get around the stress of a low-energy lifestyle. So far the only source of energy I've referred to for life is respiration. But fermentation can pull the roller coaster cars up the hill just as well as respiration can. It's just a much smaller hill. Like respiration, fermentation is a redox reaction, but instead of oxidizing one chemical to reduce another chemical, a single chemical is broken into multiple molecules, some of which are more reduced and some of which are more oxidized. Instead of oxidizing food completely to carbon dioxide, wringing every last bit of energy out of it like respiration does, fermentation leaves high-energy leftovers like acetate and ethanol. Fermentation takes one slice out of the pie and puts the rest back into the fridge for others to eat later. It's very polite. Because of this restrained eating, fermentation ends up being one of the lowest-energy processes known to support life.

Many intraterrestrials can ferment, but mostly we know about fermentation from the oxygen-free parts of the surface world, like cow rumen and beer vats. Fermentation breaks down tough, complex organic matter into small pieces that even delicate flowers such as ourselves can digest, as it produces things like beer, wine, yogurt, injera, kimchi, bread, natto, cheese, sauerkraut, and chocolate. Humans have been harnessing the fermentative power of microbes for thousands of years to make our food tasty, nutritious, alcoholic, and well-preserved.

Or, as with our chemical reaction and taxi driver example in the previous chapter, one could view this relationship from the microbes' perspective. Perhaps microbes evolved to leave more nutritious food for us humans to eat, enticing us to keep the fermentation going—to make sure we keep placing them in special jars full of their favorite food while they flourish inside. Many folks who have started making sourdough bread for fun quickly find themselves baking more bread just to keep their starter culture going, even if they don't particularly need another loaf. So who's in charge? The microbes or the bread makers?

The ΔG for fermentation is extremely low,[6] so why is something with such terrible energetics so successful? The secret is that even if you're stuck with a low ΔG lifestyle, you can still be a high roller. You just have to take *time* into account. The trick is to run your low-energy-yielding chemical reaction really fast, which is what the surface-world fermenters do. To illustrate the point, let's consider it in human terms. Celery does not have as many calories as a hamburger or a candy bar. If you replaced the higher-calorie foods you normally eat each day with an equivalent amount of celery (say, 20 or so stalks), you would soon die of starvation. But if you ate a gigantic pile of celery per day, say 250 stalks, then, thermodynamically speaking, you'd be fine, although you'd be very sad. Like a human enduring a celery-only diet, these fermentative organisms burn through a lot of substrates. The fermenters that we use in food production have low energy but high power, since power is the rate of energy delivery.

But intraterrestrials are not getting shovelfuls of malted barley in a beer vat. How much power is available for subsurface life? Are they the sourdough of the underground, growing rapidly even though they receive their energy in small packages? We already understand how to calculate ΔG, so to calculate power, we just need to know how fast the reactions happen. To do this for the intraterrestrials, we can use a nice quirk of marine sediments, which is that they are often laid down at a constant rate over time. When we lay a core of mud on its side, we can stretch a measuring tape along its length and convert the length to time. Therefore, when we measure the change in concentrations of chemicals such as sulfate along the length of these cores, we can calculate how quickly they're consumed.* Multiplying this rate by ΔG tells us the *power* that's available to the intraterrestrials.

The result is an astoundingly low value. Even the lowest-energy laboratory culture still requires thousands of times more power than what's available to these intraterrestrials.[7] In this way, life in the subsurface has drastically changed our understanding of how much power is necessary for life. The implications are that even though the deep subseafloor is one of the largest ecosystems on Earth, *hardly any of the microbes who live there are actually growing.* They have 0.00001 percent of the power that supports all other known types of cell growth on Earth, so even performing a single cell division is impossible.[†]

---

* You also have to do some other things, like subtract movements of sulfate from diffusion, compaction, and advection, but for our purposes here, we can skip over these details.

† If this fact is wigging you out, keep reading. I'll wrestle with the evolutionary implications of living for millions of years in the next chapter.

This is not something I thought I would spend my life studying— a giant ecosystem where nothing grows. Yet here I am. It's fun to work on something so completely outside my ordinary world. For instance, we know from *E. coli* that organisms breathe oxygen until it is depleted, then they use nitrate until it runs out and then, finally, they ferment. *E. coli* also uses the best food first, employing "catabolic repression" to avoid eating the low-energy food until the good stuff runs out. This is how the world of microbial energetics is presented in most textbooks. But the intraterrestrials show us that *E. coli*'s approach to eating and breathing should not be treated as a universal law of microbial life; they're just guidelines. The intraterrestrials ignore these guidelines and go straight for the low-energy stuff, allowing them to compete with other organisms using their nearly supernatural control over thermodynamics and time.

It seems that biology has universal laws after all. They're just far more relaxed than our intuition would suggest. By playing within the boundaries of these universal laws, life has developed some extraordinary ways to alter their thermodynamic landscape in ways that are foreign to our daily experience.

One of the strangest implications of the intraterrestrials' thermodynamic landscape is that, according to the calculations I've described above, most of the cells in the vast global marine subsurface biosphere never grow at all. This is a bold statement. How can an organism prefer to live somewhere it can't grow?

How does Darwinian evolution even work in a place where there are no progeny and therefore no natural selection? To answer these questions, we need to resist the knee-jerk tendency to reject ultralong lifespans as unreasonable. Like the microbes who dive so deep down in their thermodynamic hole that they pop out on the other side and reverse their metabolism, we have to sit with the discomfort of running afoul of Darwinian evolution long enough to understand the implications of long-term nongrowth. Understanding how evolution works in the subsurface will require that we reconsider some assumptions about lifespan that we probably didn't even know we had. Can we even imagine a world where an individual lives for millions of years?

PART III

HOW DO INTRATERRESTRIALS AFFECT OUR CONCEPTIONS OF OURSELVES?

8

IMMORTAL MICROBES

Bogue Sound, Carteret County, NC, Summer 1987

Cicadas bleated in my ears and my laundry basket brushed against my hip as I crawled through water so tepid its boundaries merged with the thick air. Salt encrusted my skin and mud clung to my knees. My father had tied the laundry basket to my waist with some algae-fied rope from his skiff. He, my brother, and I were scallop hunting in Bogue Sound, a lagoonal estuary in North Carolina.

Scallops are hand-sized bivalves with fluted shells. They're like clams, but more gregarious. Instead of burrowing into the muck and retracting their siphon and foot when disturbed, scallops sit at the surface with their shells wide open—staring at you with two rows of sapphire blue eyes and a fringe of little feelers. They're beautiful, and when I was a nine-year-old kid, I loved holding them near my brother's face, half in and half out of the water. When they snapped their eye-bejeweled mouths shut, I could shoot water into his face. Splashing him by hand would've been easier, but a scallop adds style. The laundry basket was for collecting the scallops we pulled up by hand, keeping them submerged during our hunt. Later that evening, we would eat them in my dad's one-room efficiency apartment, sautéed in a mess of butter. And yes, we used those baskets later for laundry. Our finances did not support a lavish, two-sets-of-laundry-baskets lifestyle.

Every time I scanned the shallow water for scallops, a whole new world emerged. Brown, red, and green algae alternated with sea grass and sargassum seaweed, forming dark shapes under the water. Stingrays flapped around, presumably secure in their ability to evade or stab predators like me. Hermit crabs decided the coast was clear, poked their legs out from whoever's shell they had stolen most recently, and tiptoed their contraband along the seafloor. Occasionally, an area of mud the size of my two outstretched hands would ripple at the edges and skitter away, revealing itself to be a flounder. A flounder is a flat fish that lies on only one of its sides at the seafloor. As it reaches adulthood, one eye migrates to join the other on the upward-facing side. Flounders sound like something out of a sci-fi novel, but they're normal as rain in eastern North Carolina. And they're delicious fried.

In many ways, Carteret County in eastern North Carolina was my first field site. Here we are surrounded by water—so much so that during hurricane season we are often consumed by it. Interaction with the ocean and all the things in it is woven into the culture. Most of our parents made a living in the seafood or tourism industries, and I had plenty of contact with fishermen, seafood processors, oceanographers, marine archaeologists, and boat builders. There was even one self-proclaimed pirate named "Sinbad" who was never out of character. After church on Sundays, many people would jump in their boats and go to Shackleford Banks, a beautiful barrier island inhabited only by wild ponies, pelicans, and semidrunk sunbathers. Shackleford is also a gold mine for dead things washed up from the sea. My brother and I loved to poke sticks at rotting corpses of jellyfish, fish, sharks, and dolphins. Once, a dead loggerhead sea turtle washed up, so an enterprising park ranger* buried it, exhumed the clean bones a year later, and donated it to my high school. My science teacher created an after-school club called Tetrapods, in which we painstakingly epoxied the skeleton back together.

Much of the area, called the Crystal Coast, is protected as the Cape Lookout National Seashore, the Rachel Carson Nature Preserve, or Fort Macon State Park. I volunteered with the park service as a kid, riding all-terrain vehicles across the white sand beaches scanning for sea turtle tracks so we could mark and protect any eggs. I also did interpretive presentations at Fort Macon State Park in Atlantic Beach. If you went for a relaxing beach day there in 1993, you were almost certainly accosted by a bushy-browed teenager wearing a park service vest, yammering about sea turtle egg-laying habits, and making you touch a live snake from her terrarium. If you were one of these people, I hope you learned something interesting, and I didn't scare you too much with the snake.

These childhood adventures feel like eons ago. For some intraterrestrials, however, this decades-long interval would be no more than the blink of an eye (or rather, the crank of a flagellum). As we saw in the last chapter, many intraterrestrials live in environments where they don't get enough power to produce new cells. As a result, individual cells might live for thousands, or perhaps even millions, of years. The oldest multicellular organisms are sequoia trees, which can live to be a thousand years old. However, the cells that make up sequoias

---

* Keith Rittmaster is his name. He now has a whole collection of bones of beached whales, dolphins, and turtles that you can visit at Bonehenge, through the North Carolina Maritime Museum in Beaufort, NC.

are much younger, because they, like all big life up on the surface of Earth, must continually renew themselves to keep pace with the fast-growing world around them. But must all life be like this?

In this chapter, I will explain the reason why I think we've overlooked the possibility of ultraslow life thus far. And then I'll address the crucial question that I've been ignoring up until now: Is there direct evidence that living beings can survive for thousands of years or longer? Finally, I'll tackle the sticky issue of how an organism evolves to stop reproducing for thousands of years or longer. To wrap our minds around this, we'll need to throw some of our implicit assumptions about natural selection directly into the trash bin.

Why Do We Assume Bacteria Grow Fast?

Microbes, at the outset, do not seem like good candidates for the slowest life on Earth. We know that bacteria grow quickly. If I wake up with a tickle in my throat, I get a feeling of dread because I know that by the evening, I'm going to have a full-blown case of strep throat—that's a lot of bacterial cell divisions in just a few hours. Some microbes grow so fast that they quadruple their genome instead of doubling it, which allows them to whip out four new cells at once. When Richard Lenski, a scientist at Michigan State University, decided to do a long-term study on a fast-growing organism so he could see evolution in real time, he picked *E. coli*, which doubles every twenty minutes when you grow it in a laboratory. In fact, it grows so fast that Lenski has been able to watch it evolve during his own lifespan. If there's one piece of common wisdom about bacteria, it's that they grow like wildfire.

However, *E. coli* and other popular fast-growing microbes don't live in subsurface environments, as far as I can tell. And, as I explained in the previous chapter, in many places, such as deep marine sediments, the power that's available from surrounding chemicals is only sufficient to replace broken body parts, not divide into two new daughter cells, so long-term metabolically active dormancy is the only option. But how long can a cell live like this? From a purely theoretical standpoint, there's simply no limit on how long a single organism could live if it slowly replaces its broken bits over time. So counter to that common wisdom about bacteria, maybe they're not all fast growers. This brings up a real conundrum. On one hand, if anything like immortality were common, then we would be surrounded by beings that were born sometime around the Big Bang, which is not the case. But on the other hand, these intraterrestrials seem like they could live forever.

Luckily, there's a lot of temporal real estate between a twenty-minute doubling time and immortality. What if the intraterrestrials live for five hundred thousand years or a million years? The oldest modern-day sediments that haven't yet turned completely to rock are about a hundred million years old, so this is an upper limit on the age of an individual cell in marine sediments. Older rocks could have older cells, as long as the rock was never buried to sterilizing temperatures over the course of its journey through the body of our tectonically active planet. It's impressive to imagine the events that occurred during the lifetimes of such individuals: the building of the pyramids, the development of agriculture, many ice ages. Theoretically, when I scoop samples out of deep sediments, I could be touching microbes that have been living and breathing continuously since before humans were even a species. If they could talk, imagine what they could tell us about our own evolution!

Perhaps our assumption that bacteria grow quickly is an artifact of how we study them. As microbiologists, we prefer to run experiments on microbes that grow fast. Twenty years ago, when I was running experiments to see how deep-sea microbes handle multiple stressors,[1] I certainly couldn't wait four weeks for each experiment. Our fast-growing cultures gave us answers overnight, which allowed us to tweak the conditions and run the experiment again quickly. If we had to wait weeks or months to see the results of every little experimental adjustment, we never would have been able to complete our work. Ergo, the only microbes we have good research about are the ones that can grow fast. So now that we know of the existence of huge, deep branches on the tree of life that have thus far resisted growth in pure cultures, maybe the universality of fast growth should be reconsidered. The necessity of laboratory expediency may have given us the incorrect impression that most microbial life on Earth grows fast.

From laboratory experience, we know, for instance, that when microbes are deprived of what they need to grow, many of them enter some form of dormancy (e.g., by ceasing cell division, slowing metabolic rate, or forming an endospore) until they can grow again.[2] In winter, soil microbes wait for warmer temperatures. Microbes that cause diseases like tuberculosis can stay dormant in our bodies for years, waiting for our immune systems to stop bombarding them.[3] However, as with the rest of microbiology, we have learned about dormancy by studying fast-growing microbes that have temporary flings with nongrowth. If there are organisms on Earth that are in permanent slow-growth states, our current methods of knowledge production would be inadequate to teach us anything about them, or about what we can do to coax them to move between slow-growth and no-growth states. That's probably why we've never noticed them until now.

Evidence for Ultraslow Growth

There is steadily growing evidence that the microbial world may be a lot slower than we think. It has long been recognized that more cultures can be obtained from natural samples if they are incubated for longer.[4] Cultured microbes that are highly abundant in seawater like *Nitrosopumilus* sp., *Pelagibacter ubique*, and *Prochlorococcus* sp. have doubling times of a day or longer, which means that it can take more than a month to run a single experiment on them.[5] Some marine sediment intraterrestrials are even slower, with Atribacteria doubling over five days, Loki-archaeota doubling over fourteen to twenty-five days, and uncultured members of the Methanosarcinales, called ANME-2, doubling over seven months.[6] These cultures grow so slowly that running experiments on them is nearly impossible on human timescales. And these are just the few microbes that have been grown in a laboratory. What if there are microbes that are even slower? We would miss them entirely.

In fact, I will venture a guess that slow growth may also be important for our normal fast-growing cultures when they're out in the environment. Even if a species grows to a robust population size overnight in a laboratory, it almost certainly doesn't grow that fast in its natural habitat. If microbes out in nature grew as quickly as even the slowest laboratory cultures, they would double the entire mass of the planet in just a few years. Certainly, these microbes act differently in a slow or non-growth state than they do when they're running full throttle in a laboratory. Even *E. coli* is a slow grower when it's floating around, contaminating a drinking water reservoir. Fast growth is an aberration, even in those species capable of moonlighting as fast growers.

These limitations force us to get creative about how we study microbes that are dormant for many years. To us, they look like they're doing nothing. As an analogy, over a geological timescale, the California coastline is a constantly churning mass of rocks, but on our human timescale, it is stable enough to build a house on that we can pass to our grandchildren. These houses must be sound enough to withstand the occasional earthquake, but they will not survive the reorientations of land as they are spun, submerged, and exhumed over the course of a few million years. To think like an intraterrestrial, we have to grapple with some uncomfortable timescales.

The importance of timescales for biological research came into stark focus for me in 2003, when I was tramping around the hot springs of Yellowstone National Park. I was a student in the Agouron Geobiology Summer Course at the University of Southern California. We were learning about the microbes that live in the steamy water fizzling up from the giant volcanic caldera that lurks beneath the park. I squatted next to a boiling hot spring and watched as a spider skittered across its lethal surface, dancing along water that would have killed me for sure. I wondered if the spider had super-heat-insensitive hairs on its feet, creating a cooler boundary layer between the water and its body.* Or did the spider move quickly enough that only a small amount of heat was transferred into each foot before it was raised again into the cool air? The spider was twitching like mad, so I favored the second hypothesis.

---

* There's a deep-sea worm that does something like this, called *Alvinella pompejana*. It immerses itself in 100°C water in deep-sea hydrothermal vents, protected by hairs that create a cooler boundary layer of protection from the dangerous fluids.

Just as I was planning to quit microbiology and make a career out of studying this fascinating spider, its deranged twitching slowed. Then it stopped. Then the spider curled up and died. I almost laughed out loud at my misunderstanding. I wasn't watching a spider with extraordinary adaptations. I was watching the unfortunate demise of a spider in a hot spring. The problem was that my timescale was off. I assumed that the few minutes I had observed the spider were representative of its normal life. If I hadn't had enough time to sit still and watch it for a while, perhaps I would have seen the spider scampering for a few moments (in reality, it was struggling for dear life) and assumed I had the whole story. It was only by watching the spider for many minutes on end that I saw how wrong I was.

Let's do a better job than I did in Yellowstone and get the timescale right for marine sediments. In the previous chapter, I noted that the power delivery to intraterrestrials in marine sediment is many orders of magnitude lower than that required to support laboratory cultures in a nongrowing stasis.[7] These models predict that the whole microbial population turns over, or refreshes, by replacing broken parts, not by making new cells, somewhere between every few years and tens of years in most of the mud underneath all the world's oceans.[8] This alone is good evidence that these microbes hang out in suspended animation for hundreds, thousands, possibly even millions of years without ever growing and making new cells. Nevertheless, this is a radical conclusion, so we need some direct evidence to support it.

Elizabeth Trembath-Reichert at Arizona State University and Julie Huber* at the Woods Hole Oceanographic Institution sampled fluids containing microbial communities buried three hundred meters deep into solid oceanic crust that itself was over

---

* In a previous chapter, I described how Julie pioneered the use of 454 high-throughput DNA sequencing to discover that life was much more diverse than we thought.

three miles deep into the ocean. Just getting your hands on samples like these is an accomplishment on its own. Then they added water labeled with various chemical tracers to see how quickly the microbes took the tracers up into their cells to make new biomass.[9] Amazingly, they saw the cells take up the chemicals, although it was achingly slow—most cells took up only about 0.001 to 0.1 femtogram (fg) of carbon per cell per day. Given that there are 19–31 fg carbon per cell in marine sediments,[10] this means that it took somewhere between a half a year to eighty-four years for each cell to turn over its biomass—again, without actually making a new daughter cell. This direct evidence falls squarely within the predicted turnover times based on thermodynamics above.

We have other direct evidence that marine sediment cells are in long-term stasis. Since we know how fast the layers accumulate vertically, we know that deeper cells (or their lineage, if they've been growing) have been buried for longer. If these microbes are dividing into new cells, rather than just completing a part-by-part whole-body replacement, they should show genetic modifications over many hundreds of generations that we would be able to observe in the upper few meters of sediment. But they don't. The deeper microbes lack the kind of genetic changes we would have expected if they were undergoing cell division over that period of time.[11]

We know these cells are alive because we can see the effects of their respiration on the chemicals around them and we can see that they are intact and not degraded. So they must be refreshing their cellular biomass, but they are doing it by gradually replacing their parts, lipid by lipid, nucleotide by nucleotide, such that over roughly half a century, all the molecules will have been replaced. They are not actually dividing and making new daughter cells; instead, individual cells survive for thousands or even hundreds of thousands of years. So direct evidence currently supports the theory based on thermodynamics: living cells in deeply buried marine sediments survive for thousands of years or even longer without dividing into new cells. It's extraordinary, but, at least for the moment, it appears to be true. This means we need to address the following difficult question: How does a species that doesn't produce progeny for thousands of years or longer even evolve, given the importance of progeny in the process of evolution?

How Does One Evolve to Stop Growing for Thousands of Years?

To answer this tough evolutionary question, first we have to think about what these organisms would experience in their lifetimes. These slow organisms wouldn't be concerned about the length of a day. They're buried so deep that they can't detect the sun anyway. They probably wouldn't even notice a change in season. However, they might care about other, longer geological rhythms: the opening and closing of oceanic basins through plate tectonics, the formation and subsidence of new island chains, or new fluid flows brought on by slow formation of cracks in Earth's crust. The biology I was taught in school considered these events to be evolutionary drivers for a *species*, not an *individual*. For instance, Darwin's finches evolved new beak shapes because they had been isolated on an island with a particular shape of seed to eat. This evolution happened over the geological timescale of island formation, but it occurred in a species lineage, not in an individual bird.

We know, however, that individuals are also capable of changing along with the rhythms of their environment. An individual arctic fox's fur changes from white to brown when the snow melts every spring. Many people (though sadly, not me) wake up at the same time each morning without the aid of an alarm. Daily and yearly rhythms seem like reasonable things for a person or an animal to keep track of. Ice ages, less so. Anticipating changes over longer timescales seems ridiculous. It would be silly to suggest that an individual finch would have evolved the ability to swim because it had an innate anticipation that its island would subside into the sea in a hundred thousand years. Or that a beetle in the Gobi Desert could only reproduce when it ate an Amazon rainforest seed because it was born millions of years ago when South America and Africa were nestled into each other, and its DNA instructed it to reproduce when the tectonic gap closed again. These scenarios make no sense for animals, but they may be reasonable for the intraterrestrials. An individual that lives for a million years might be evolutionarily predisposed to count on something as slow as island subsidence in the same way that we are evolutionarily predisposed to wait for the sun to rise tomorrow. To fully understand intraterrestrials, we may have to rethink what qualifies as an evolutionary cue.

The fact that living cells likely exist in a nongrowth state for very long timescales raises two important questions. Can a microbe be adapted to *avoid cell division* for thousands of years or longer, rather than having it just happen by accident? And, if so, how does evolution work for an organism that seemingly never produces offspring?

Let's tackle that first question by stating it this way, in order to help us place this finding in the context of Darwinian evolution. Are these microbes *evolutionarily adapted* to hang out in this undead, dormant state for thousands or millions of years, or do they just persist because cells don't need any special adaptations to stay alive for so long? To me, living for hundreds of thousands of years seems unlikely to happen without adaptation. Too many physiological changes are required to support this lifestyle for it to be a side effect of a "normal" fast-paced life. Furthermore, if this lifestyle is accidental, then their main growth-supporting lives must occur in some other environment. But we rarely see the types of microbes we find in the subseafloor elsewhere. It's not as if they were normal seawater microbes happily swimming around, dividing, and growing when they fell to the seafloor and forgot to die. On the contrary, most of this highly diverse group of microbes seem to exist only in marine sediments.[12] Given this, they may be just as selected-for in marine sediments as parrots are in a rainforest. Indeed, we find that at increasing depths in marine sediments, microbes make enzymes with a higher specificity for the type of substrates that are available in the subsurface, suggesting that they are specially adapted for this environment.[13] Subsurface microbes also have adaptations that enable ultraslow metabolisms and cell divisions.[14] This suggests that they are somehow evolutionarily poised to be in a long-term nongrowing state.

But here we have a problem. According to Darwin's theory of natural selection, these cells must grow and make new progeny to evolve. Natural selection works because, during reproduction, organisms experience mutations. And when an organism has a mutation that is beneficial, the mutation increases the organism's fitness, so the organism's progeny outcompete those of the nonmutated organisms, resulting in more progeny that have the mutation. These further generations continue to do better than the nonmutated lineages, and eventually the mutation spreads throughout the population. Voilà, adaptation has occurred through natural selection. But how can we even think about Darwinian evolution in populations that don't reproduce? How can you become adapted to *not* have babies?

I don't think Darwin had nongrowth in mind when he described survival of the fittest.

Luckily, we have a good model in short-term seasonal dormancy. Here dormancy during winter has an evolutionary advantage because the dormant organisms have larger populations remaining once conditions are ripe for growth again in the spring. These organisms thus have a head start on other organisms and can pass their dormancy genes along to a larger population of progeny in the spring and summer. This is textbook Darwinian natural selection. Let's extend that model to dormancy that lasts for thousands of years in marine sediment. We have to think of an event that intraterrestrials could possibly be waiting for that would pull them out of dormancy when they're buried hundreds of meters deep in Earth's crust. If we encounter a dormant microbe in soil in winter, we can presume that it's waiting to start growing again in summer. What is the equivalent situation for a deeply buried marine sediment organism that is dormant for thousands to millions of years?

Let's do a thought experiment to jailbreak our brains from our implicit assumptions about lifespan. Imagine human lives only lasted about twenty-four hours. You'd be born at midnight, rebel against your parents at breakfast, settle down and have babies just before lunch, and pick up fishing as a retirement hobby around dinnertime. By midnight, your loved ones, who themselves were only born a few hours ago, would huddle close and hold your hand as you'd pass away peacefully at the ripe old age of a day. If everyone did that, hundreds of human generations would come and go within a single winter. Throughout that time span, which would represent a significant chunk of human history, the deciduous trees would remain brown and lifeless. The permanent deadness of trees would be taken as an undisputed fact and scientists like me would probably write grants to understand whether or not trees are alive, given that they don't seem to grow or make progeny. Of course, if you stretched back far enough, humans would have been present for the fall or even summer, but that might have been so many generations back that a stable form of writing had yet to be invented. We hundred-year-lifespan humans know that trees are just waiting to take advantage of the summer sun. But the day-lifespan humans would be stumped.

When we think about life in the subsurface, are we like day-lifespan humans contemplating a tree? Are long-lived intraterrestrials waiting for wake-up cues we don't recognize because our lives are too short to see them? What is even the point of living for hundreds of thousands of years anyway? There must be *some* reason these intraterrestrials stick around so long.

There is evidence that long-term dormancy has a selective advantage. When you let the laboratory workhorse *E. coli* sit around with no food for months or even years, many of the cells enter a state of long-term dormancy where they are alive and metabolizing, but they're not growing nearly as quickly as they do when you feed them. If you mix these next-to-dead *E. coli* with a fresh batch of fast-growing *E. coli* and starve them both, the old geezers beat the living daylights out of the sweet little young'uns. This growth advantage in stationary phase (GASP)[15] may be the secret to why intraterrestrials live so long. Maybe they're waiting for something that only happens thousands of years later so they can be the ones to take advantage of the new situation. They might act as monks, accustomed to deprivation while the gluttons die around them.

So what are these microbe-monks waiting to wake up for? Seasonal cycles are way too fast. The only things slow enough are geological processes. For instance, island subsidence, floods, drought, or storms often occur on hundred- to thousand-year cycles. Submarine landslides, earthquakes, tsunamis, and volcanic eruptions might shift materials around on even longer timescales, exposing intraterrestrials to new food sources that coax them out of dormancy after hundreds of thousands of years. It seems odd to say that a microbe is adapted to wait for something as infrequent as a volcanic eruption, but Earth's history shows that you can rely on volcanic eruptions, as long as you've got time to wait for them.

If we really let our imagination run wild, individual microbes might be adapted to events with even longer periods like glacial cycles, which shift every thirty thousand years or so. Or the slow movement of tectonic plates. As new seafloor pops up in mid-ocean ridges, the existing seafloor is constantly being pushed away from the middle of the ocean, like a person standing on a moving walkway at an airport. The seafloor eventually jams into a continent in the slowest-motion train wreck ever.

Some of the sediments and the intraterrestrials that live in them will get dragged on the subducting plate down to eventually be crushed at temperatures and pressures that kill all life as we know it. Even for extremophiles, being dragged all the way down to the mantle would definitely be an evolutionary dead end. However, some of the sediments that are in the early stages of being subducted under continental plates might be returned through cracks and fissures that open in the overriding plate. During this collision, some of the seafloor sediments are shoved upward in accretionary prisms and the attendant faults created by earthquakes or other plate deformations. Could all this piling up, faulting, and burbling up to the surface be what the intraterrestrials are waiting for?

Let's think through the implications. This would mean that the individual microbial cells that we pull up in our drilling ships that appear to be dormant are just waiting patiently for the ultraslow movement of the plates to squish them into a continent, where they have a chance of resurfacing and recommencing growth. The evolutionary payoff for waiting for millions of years in deep marine sediments would be to return to the upper seafloor again where the food is more nutritious, at which point the microbe would pass its genes along to future generations. Like any standard Darwinian natural selection, the individuals that have the best adaptations to being dormant for millions of years would have a growth advantage once they arrive back to the surface, ensuring that those adaptations become stable in the communities. Is getting tossed back up into surface sediments an intraterrestrial's version of summer?

Evidence for an Intraterrestrial "Summer"

For these theories to make sense, we need evidence that the species of microbes that we normally find in a nongrowth state actually do grow and produce new cells in shallow marine sediments. In principle, shallow sediments are a plausible place for such growth, as the nutrients at the seafloor are the freshest they'll ever be.* But it's hard to prove that microbes grow in these shallow sediments because there's no easy way to measure microbial growth rates in nature. One way of measuring the growth of slow-growing organisms in natural samples is to adopt the method that Trembath-Reichert and Huber used to work in deep basaltic crust, which I mentioned previously. However, this method still involves adding tracer to the system.

---

*I have to credit Andreas Schramm at Aarhus University for first putting this notion in my head when we were chatting at the American Society of Microbiology meeting in 2014. When he said that microbes *must* grow in shallow sediments, it was like a lightbulb went on in my brain!

I wanted to see intraterrestrials growing without disturbing them at all.

I recently thought of a different, more passive way to measure the growth of intraterrestrials in shallow sediments. I thought, if we can measure metabolic rates by changes in substrates over the depth layers in marine sediment, why can't we use the same method to look for microbial growth rates? If they grow slowly enough that their population increases at a similar rate as their burial, then maybe I could calculate the growth rate by measuring the increase in the number of microbes in different depth layers.

To put this another way, imagine you're a microbe sitting in sediment. You don't know you're slowly being buried, but you know there's enough food around you that you can grow right now. But you grow very slowly, so you only make a new cell once every three months. If the sedimentation rate is a steady four years per vertical centimeter, then when you are buried 1 cm deep, you will have made about one hundred thousand new cells of yourself, minus any that have died in the meantime, while cells from your species are also constantly being buried above at roughly the same rate that they were when you were buried.* By comparing how many cells exist at your depth layer after growing for four years to the number of cells of your species 1 cm above you, you can calculate the net growth rate for your species.

Sounds simple, right? Unfortunately, there's an alternate explanation for the observation that there are roughly one hundred thousand times as many cells of your species at 1 cm deep than there are at the surface. It could be that your species grows

---

* This assumption is definitely violated in some places—for instance, if there's been a recent mudslide—but we can account for this by looking at the changes in geochemical tracers that are not affected by biology.

faster than one cell division per three months, but something is different about the deeper sediments that allows them to accommodate more of your cells. According to this picture, your species is in steady state all the time, with growth and death balancing each other to achieve the maximum number of cells that can be supported by the environment (called the environmental carrying capacity). If this is the case, then an increase in number of cells with depth is not a growth rate, but an increase in total carrying capacity of the sediments. On some level, however, it doesn't matter whether the rate of increase in cells with depth represents a growth rate or an increased habitat carrying capacity. Either way, finding that there is an increase in cells of a particular type of microbe with depth is evidence that certain species are growing, even though the other species are dying. So to catch intraterrestrials in the act of growing, I just needed to figure out what shallow marine sediments to collect.

I have discussed some exotic fieldwork over the course of this book: summitting volcanoes in the Andes, evading polar bears in the Arctic, diving to skull-crushing depths in submersibles. But microbes have taught me that, if you pay close enough attention, what may seem mundane is actually extraordinary. In eastern North Carolina, for instance, oxygen-rich air, grocery stores, bathrooms, and gas stations are plentiful, the ground is unlikely to explode, and nothing is going to actively hunt you. But I've spent more time chasing the edge of a lightning storm while speeding in a motorboat over rough seas or knocking on the side of my little Sunfish sailboat to call over dolphins as they leap and spray sun-sparkles all around me in eastern North Carolina than I ever have anywhere else. I can say with some authority that my hometown of Beaufort, North Carolina, and the surrounding area are pure magic.

What is also magic about eastern North Carolina is that it contains nicely layered marine sediments that just happen to contain many types of the same intraterrestrials that we find in the deep sea. This makes it the perfect place to see if we can catch these microbes growing in shallow coastal sediments, where the food is as good as it'll ever get for them. In May 2016, my lab members and I drove to my hometown, collected some mud, and sliced it into fine layers. We found that although the total number of microbes decreased with depth, some populations of microbes *increased* with depth, with doubling times of five to twenty-five years.[16] And the types of microbes that were growing were the same groups of intraterrestrials that we find in this long-term nongrowth state in the deep subsurface. We also found that normal seawater microbes, on the other hand, died off pretty quickly with depth in the sediments.

Of course, this was just one study. But for now, at least, our work appears to support the idea that the "summer" for deep marine sediment microbes—the time when they make progeny for natural selection—might be right near the surface of marine sediments. If that's the case, the point of their ultra-long-term dormancy might truly be that they're waiting on a rare geological event such as subduction, eruptions, landslides, or earthquakes to push them back up to their shallow "growth zone."

The Aeonophiles

Despite what we know from our lived experience, the intraterrestrials are showing us that Earthlings can live for many thousands of years or longer. And some of them* may sync up

---

*Notice I've only been talking about nongrowth states in the intraterrestrials that live in marine sediments, even though there are many other types of subsurface

their growth cycles with ultraslow geological movements to get ahead in life. Super cool. In my opinion, these are fundamental discoveries about the nature of life on Earth.* In fact, I believe that the discovery of ultra-long-lived creatures is up there with the discovery of thermophiles, microbes that thrive at temperatures above the boiling point of water. When thermophiles were discovered, it blew open our understanding of where life might exist in the universe. It also launched an entire industry focused on heat-tolerant enzymes and organisms. I foresee a similar seismic shift from the discovery of long-lived intraterrestrials. The existence of such organisms greatly expands the window of time during which we can look for biomarkers in the universe. As for industrial technologies, stable microbes make stable enzymes.[17] And stable enzymes make money. Longer-lasting, tougher enzymes are good for the bottom line in a biotechnology company that depends on them. But if we're going to turn these tiny long-lived oddities into a legitimate subject of study, we're going to have to give them a name.

Thermophiles like high temperature, psychrophiles like low temperature, acidophiles like acid, barophiles (also called piezophiles) like pressure, halophiles like salt, and alkaliphiles like alkalinity. The thing that the long-lived ultradormancy microbes like is *time*, which serves as a resource that microbes can exploit to access new habitats. They are able to wait for

---

environments. Intraterrestrials are probably just as slow in deep ancient terrestrial aquifers, it's just harder to measure their growth rates because we don't have the luxury of nicely layered sediments to record the passage of time. Other subsurface environments may also have ultra-long-lived organisms, or they may have faster-growing microbes, fed by some other source of substrates.

* Lest this appear entirely self-congratulatory, I want to be clear that my own research is only a drop in the bucket of scientific discoveries in this field.

resource-replenishment events that are farther into the future than any other species is able to hold out for. To signify their love for long stretches of time, I suggest the name *aeonophiles*, from the Greek "aeon," meaning "an indefinite and very long period of time." ("Aeon" is the origin of the American English word "eon" and is pronounced the same way.) Geological events may be as consequential to an aeonophile as daybreak is to a songbird.

———————————

To put these aeonophiles into proper context, let's examine them with a classic ecological framework that distinguishes between *r* strategists, who "live fast and die young," and *K* strategists, who reproduce later and live longer.[18] Mosquitoes are *r* strategists, living for about a day but making so many babies that they're guaranteed to still be annoying long after humans are extinct. Sequoia trees are *K* strategists. They like to settle in for a few centuries before they start considering reproducing. In a way, aeonophiles are extreme *K* strategists. The only problem with applying these *r* and *K* terms to microbes is that this framework was developed for big things like flies and trees, so the theory makes predictions about offspring, sex, and body size that do not perfectly translate to microbes. But if we can roughly shoehorn microbes into this *r* versus *K* ecological theory, then the novelty of viewing time as a microbial resource does not signify a new ecological paradigm, but rather a new ecological niche. Aeonophily is a niche that distinguishes this group from organisms that operate in an ecological world where everything dies in a hundred years or less, just as thermophily distinguishes organisms from those that require temperatures at or below 37°C. In the case of aeonophiles, their ecological "niche" is less a place than a temporal dimension.

As we continue to learn about aeonophiles, I predict they're going to continue smashing our preconceived notions on how life is supposed to work. They may even give us lenses through which we can finally peer back in time and answer the most fundamental questions of our own existence: *What/When/ Where is the origin of life*? Or even that really tough fourth question: *Why* did life happen?

9

RETHINKING OUR BEGINNINGS

INTRATERRESTRIALS GIVE us a new perspective on the origin of life because they give us a new perspective on the nature of life itself. They've shown us that you can toss hundred-year lifespans to the curb if it suits your evolutionary lineage, that shimmying to the edge of thermodynamic limits is just fine if you have the right enzymes, and that nearly every rock on Earth can provide chemistry to support life. These facts bring our conception of life much closer to our conception of geology, which is what needs to happen if we're ever going to figure out how life arose from what was previously only rocks, minerals, gases, and fluids. Luckily, the idea of merging biology and geology is not new. In 1926, Vladimir Vernadsky (1863–1949) coined the term "biosphere" to talk about them as a unified system.[1]

In this chapter and the next, I'll take my best crack at using this unified biosphere system to answer the three fundamental questions of life: *Where, how,* and *when* did life begin? Finally, I'll inch toward addressing that intoxicating fourth question: *Why* did life begin? To do this, I will describe my visit to the type of place where life may have originated: a deep-sea hydrothermal vent. Then I'll explain how what we found there has changed our conception of our own origins forever.

Where and *How* Did Life First Form on Earth?

The questions of *where* and *how* life formed are hard to separate, as location determines materials and conditions—that is, assuming life started on Earth. According to one theory, called panspermia, life on Earth is actually of extraterrestrial origin. Mars, for example, had an ancient ocean and is close-ish to Earth, so maybe life originated there. However, while pushing the origin of life off to Mars has some interesting implications— not the least of which is that we're all aliens—it doesn't answer the question of what kind of environment supported the first life. To me, life's origin has always seemed like it should have happened somewhere deep, dark, and full of possibilities.

Thermodynamics provide a way to approach questions of where and how life originated, since they ultimately set the parameters for life. At the surface, the sun provides the most powerful energy gradient on our planet. If early life needed power in a strong burst, then life probably started at the surface. However, if it did start at the surface, it would have had to quickly evolve a way to block the sun burn, since the sun's violent power rips apart organic molecules. If, instead, nascent life needed many smaller energy gradients to play off each other, forming ladders of energy that could be linked together in different ways, then the subsurface might be our ancestral home. All those rocks in the subsurface provide nice opportunities for life to settle down and enjoy a wide range of metabolic niches, protected from the sun's damaging rays.

This perspective favors the subsurface: while sunlight and oxygen are clearly necessary to support large complex mammalian and plant life, the myriad smaller gradients that exist throughout the womblike nature of the subsurface seem like great options for the spark of early life. Moreover, the protective environment of the subsurface would have given life a few million years to develop a strategy for handling solar stress before it ventured into the light. So let's begin our search in the vast underground.

Some subsurface environments are better candidates for the origin of life than others. Marine sediments, for example, seem particularly ill-suited for the job, no matter how fascinating they may be. Most of the food available in marine sediments comes from photosynthetic leftovers, so this environment couldn't have predated life. While organic matter (aka, food) *can* sometimes be produced abiotically—asteroids, for instance, carry many types of organic matter that were made by abiotic processes—it's currently unclear whether these processes could have produced food in great enough quantities to get life cranking here on Earth. I mean, it hasn't worked out for the asteroids, as far as we know.

We can therefore narrow down the likely locations for the origin of life to subsurface environments that have their own sources of energy and can support chemolithoautotrophs. These little guys, as you may recall, use chemical energy to turn carbon dioxide, which was abundant on early Earth, into organic carbon, which can feed the masses, human and microbial alike. Once chemolithoautotrophs evolved, it's easy to imagine the rest of life evolving after them. Dead chemolithoautotroph bodies could have fed lots of different microbes, which in turn fed cyanobacteria, which enabled the evolution of plants that made enough oxygen and biomass to fuel animal evolution. Subsurface serpentinizing systems* are popular choices for where and how this process may have begun. It's hard to beat water + rock = life as a pithy origin equation.

---

* I introduced these in chapter 5, when I was discussing living in bleach.

There are also good arguments for life originating in deep-sea hydrothermal vents, where deep hot water is infused with chemicals below the seafloor. When this deep water shoots up and mixes with cold oxidized seawater,* the magic of life can happen. In addition, the heat of hydrothermal vents accelerates chemical reactions, which might have been useful for nascent life. Perhaps most convincingly, when you look at the genes that are shared among all life on Earth, they often look like they're adapted to high temperatures, hinting that maybe we all came from somewhere hot. To confirm this theory, we need to study deep-sea hydrothermal vents, and that is easier said than done. They're so hard to get to that they weren't even discovered until 1977, and therefore much about them remains mysterious.

Sinking Down to the Origin of Life

By now you know that I'm drawn to mysterious places like a moth is drawn to a flame, so it may come as no surprise that in 2008 I found myself sitting at the bottom of the ocean, this time in the Sea of Cortez (also called the Gulf of California) in Mexico, two thousand meters (more than a mile) deep, in the *Alvin* submersible, the very vehicle that was used to discover the first hydrothermal vent. *Alvin* is the only human-occupied deep submergence vehicle in the US scientific fleet—it was used to discover the wreck of the *Titanic*—and it handles extreme pressure with grace. At the time of my dive, *Alvin* could've plunged to twice the depth of our mission without a problem. Due to recent upgrades, it can now go even a mile deeper than that.

The *Alvin* launch was much like that of the *Johnson-Sea-Link II* that I described in the introduction. After working our

---

*Even before the great oxidation event, the sun would have oxidized some chemicals at the surface. So they were certainly present before photosynthesis.

way through loads of checklists and shutting the hatch, we were yanked off the ship's deck by the A-frame and then nestled into the water. I sat cross-legged, but blissfully upright, in a sphere with two-inch-thick titanium walls, mere inches from the pilot, Sean Kelley, and my advisor, Andreas Teske, each with our own porthole.

I once again became very aware of the dangers that such a dive could present. In the pre-dive briefing, Sean pointed out one danger in submersible diving that had never crossed my mind: fire. If you're hanging out with a bunch of electronic equipment in a room the diameter of a dinner table, a fire from a spark in the equipment would be catastrophic. But a submersible has an advantage over a dinner table. In a submersible you have control over every molecule of oxygen in the room. And fires need oxygen. So fighting a fire in *Alvin* is simple. You put on a breathing mask, shut off the oxygen supply to the room, and use a fire extinguisher to speed up the process. To reduce the threat further, we were breathing air with an oxygen content lower than you find at sea level. Ironically, diving in *Alvin* felt more like my high-altitude work in the Andes than scuba diving.*

As with my previous dive in the Gulf of Mexico, the bioluminescent lights were stunning as we dropped like a stone through a mile of seawater. When we released ballast and the seafloor came into view, I felt like I had gone to another world. Water shimmered all around us as the heat from the hydrothermal vents rose from the seafloor. I pressed my face close to the acrylic porthole but tried not to touch it. The pilot in charge of

---

* In the spirit of continually updating its safety features, as of 2023 all the remaining 120-volt circuits on *Alvin* have now been reduced to 24 volts, so *Alvin* now lacks the electrical potential to fuel open flame anyway. Nevertheless, it still has a low-oxygen atmosphere and secondary breathing masks just in case.

keeping the portholes pristine, Dave Walter, had warned me not to bring it back to the surface smudged. On the other side of the porthole was a seething mass of gigantic worms. Each worm was roughly as tall as me with a white body the diameter of a garden hose and a bright red head like a feather duster. The worms radiated from every hard surface in chunky bouquets. I knew that I was one of only a few hundred people who have ever seen worms like these up close, so I inched my greasy nose closer to Dave's precious acrylic window.

My job was to make sure we completed our scientific objectives, documenting our work with written notes and video from *Alvin*'s external cameras. In my spare moments, I filmed the fantastic worms as they sipped chemicals from the scalding fluids. But these animals don't use the chemicals directly. They funnel them to bacteria that live inside their bodies and these bacteria turn the chemicals into food for the worm. We humans often take probiotic supplements to keep our gut microbes healthy. These giant worms, known as Riftia, take this notion to the extreme. Over the course of evolution, they cast aside their mouths, stomachs, and butts so they could depend on the microbes for everything. Now they just have a special organ called a trophosome packed with sulfur-oxidizing microbes.

I was not, however, at the bottom of the sea to study the worms, as cool as they were. No, I was hoping to catch something much older—something stretching back to the origin of multicellular life itself. When I completed the dive and returned to the lab in Chapel Hill, North Carolina, with the hydrothermal vent samples in hand, I was stunned by what I eventually found in them. I pulled out DNA, sequenced it, and immediately encountered a dilemma. As I hammered away on my computer, trying to identify these sequences based on their similarity to known DNA sequences, one set of sequences refused to fit nicely into any category.

Stumbling into the Origins of the Eukaryotes

In chapter 4, I described how DNA sequences from the subsurface introduced new deep branches on the tree of life that we never knew about before. One of these groups, however, stood out from the rest, and this is the group that was showing up in my hydrothermal vent samples. Try as I might, I could not make a phylogenetic tree that placed these DNA sequences reliably into one of the archaea, bacteria, or eukaryotes. At the time I was doing this work, all known life on Earth fell into one of these three groups, so my first thought was that I was a terrible scientist, and I was screwing up the algorithm somehow. Instead, it turned out that I had stumbled onto an organism much stranger than I had expected.

I started combing the scientific literature to see if other people had found these strange DNA sequences too. They had been found in other deep-sea hydrothermal vents and other subsurface locations, but because this microbial group was uncultured, it did not have a proper name. Costantino Vetriani at Rutgers University named it Marine Benthic Group B (MGB-B). Fumio Inagaki at JAMSTEC named it the Deep-Sea Archaeal Group. Ken Takai at JAMSTEC named it the Marine Hydrothermal Vent Group. Katrin Knittel at the Max Planck Institute for Marine Microbiology in Germany decided to try and take its picture. Taking a microbe's picture is harder than it sounds. Sure, if you have a nicely growing culture, you can smear it across a glass slide, stain it, stick it under a microscope, and snap a picture. But if you're trying to discover a new mysterious group of microbes at the bottom of the ocean that refuse to grow in a petri dish, the task is more difficult. When you smear fresh mud across a microscope slide, the cell stain does not distinguish one type of microbe from another—a significant problem since there are *billions* of different types of microbes in each gram of sediment.

To overcome this obstacle, Katrin went to a computer and lined up the 16S rRNA gene sequences from this strange MBG-B/DSAG/MHVG against those of all other bacteria and archaea known at the time and picked out an eighteen-nucleotide section that only matched this one group. She then attached a fluorescent probe to these little eighteen-nucleotide stretches of DNA and then swished it around in the seafloor muck. Somehow these bits of DNA probes found their targets: they glommed on to the MBG-B/DSAG/MHVG cells and made them glow, giving Katrin and her colleagues the first glimpses of these mysterious beings.[2] As the scientists peered into the microscope, the first thing they noticed was that these cells were extremely small; they were only about as wide as the wavelength of blue light (550 nm). But the most important thing about them, the thing that may be extremely consequential for our own existence, is that they formed globby aggregates with other cells. In any other type of microbe, the tendency to form aggregates is not earth-shattering. But these were not normal microbes. These were not-bacteria, not-quite-archaea, and not-quite-eukaryotes. They were on the edge of the known tree of life. To understand why this is so important—why the clumping together of these specific cells made pulses race—you need to know a bit about the very deep history of our branch on the tree of life, the eukaryotes.

One of the defining features of eukaryotes is that we have microscopic organelles inside our cells called mitochondria that take electron-rich food molecules (harvested by our guts and passed along through our blood) and react them with oxygen (brought by our blood from our lungs). As I've noted in previous chapters, aerobic respiration is thermodynamically powerful, so our mitochondria use it to fuel the production of ATP and create electrical signals that nudge our muscle cells into action. But our mitochondria are not really "us"—they are ancient bacteria that for mysterious reasons ended up inside an early eukaryotic cell. Then they stayed so long they became permanent residents.

The biologist Lynn Margulis developed this endosymbiosis theory, along with the five-kingdom classification scheme. And, unlike the five-kingdom classification scheme, the endosymbiotic theory is holding up well to ongoing discoveries. If you look at a microscope image of a cell from a human, puffin, plant, or most other eukaryotes,* you will see that these mitochondria look a lot like bacteria. They have their own cell membranes, and they even retain their original bacterial DNA. That means that right now, inside the cells that we think of as ours, we are carrying bacterial DNA.† And it's similar enough to DNA from modern bacteria that we can even say what type of bacteria it originally came from. It's from a pretty run-of-the-mill group called the Alphaproteobacteria, which are common in many environments, including seawater. This bacterium somehow made it into a eukaryotic cell and became so useful that it lost its ability to grow independently of its host and reduced its genome down to just a few essential features. These leftover genes are insufficient to keep the bacteria alive on its own, so now we're stuck together. We couldn't do anything we like to do without

---

* Some microscopic eukaryotes lack mitochondria, but this appears to be because their lineages lost the mitochondria, rather than their lineages never having had mitochondria at all.

† In our guts, of course, we have a whole host of microbial visitors. But the bacterial DNA I'm talking about here is inside our human cells.

our little mitochondrial interlopers. Listen to music, form a thought, read a book, or produce body heat—they help us do it all. And the same is true for every plant and animal on the planet.

But if it really happened this way—if an ancient bacterium got inside our cells and stayed—then we would expect to see other examples of it. If something works, nature doesn't just do it once. Look at the shapes of whales and whale sharks. They both look like the fuselage of a Boeing 767, but they got there through completely separate evolutionary pathways.

Luckily, there's a second example of endosymbiosis giving rise to organelles right under our noses: plants. Just as our ability to run and jump isn't really ours, a plant's ability to photosynthesize doesn't really belong to the plant itself. Like mitochondria, chloroplasts (the parts of the plant that perform photosynthesis) are also bacterial visitors from ancient times. Chloroplasts have retained their DNA well enough that we know they hail from the cyanobacterial lineage. And, like mitochondria, chloroplasts' fates are now inextricably tied to the eukaryotic cells they inhabit. But, unlike chloroplasts, mitochondria are present in all major groups of eukaryotes, not just plants, meaning that mitochondria must have been key to the evolution of our entire branch of life.

How did this association start between early eukaryotes and the bacteria that later became mitochondria? Did an ancient eukaryotic cell eat an ancient bacterium that, instead of dying, started collaborating with its captor? Did an ancient eukaryotic cell get infected by a malicious microbe that had a change of heart and chose to live in harmony with its quarry forever? Or were they just two cells chugging along beside each other, passing food and resources back and forth long enough that they decided to make their relationship permanent? It's fun to think about these possibilities, but biology is too messy to make much headway from thought experiments alone. Data is the lifeblood of biological theories.

Fortunately, we have tons of data about the ancient Alphaproteobacteria that turned into mitochondria. *Un*fortunately, we know almost nothing about the ancient cell that was their host. As I sat in my windowless room as a graduate student just back from the Sea of Cortez, trying unsuccessfully to force computer algorithms to tell me what the heck these microbes were, I may have been looking at some of the first-known direct descendants of the host that engulfed an Alphaproteobacteria and eventually turned into humans. This is why people were so gobsmacked by these boringly named* MBG-B/DSAG/MHVG microbes. What if they retained some of the properties of that first step toward eukaryotic life on Earth?

This is also why it was so enticing when Katrin Knittel showed us that these microbes form little clusters with other cells. Maybe these microbes' penchant for huddling together led to some of them forming a relationship with Alphaproteobacteria and starting along the pathway to becoming what we are today. But it's hard to identify our "great-grandma" by just one gene and a handful of pictures. Unfortunately, that's literally all we had to go on for almost two decades. Then, in 2015, Anja Spang and Thijs Ettema at Uppsala University in Sweden made the first whole genomes of these mysterious organisms.[3] To the gratitude of scientists like myself, they gave them a name that was not an acronym: Lokiarchaeota, after the hydrothermal vent called Loki's Castle (a nod to the Norse god Loki) where they were found.

When these researchers lined Lokiarchaeota's genome up against those of other organisms, they found that they were

---

* Boring, but useful! No offense to Vetriani, Inagaki, and Takai.

surprisingly close to our lineage, the eukaryotes. The fact that they are so closely related to us suggests that they may truly be direct descendants of our ancient progenitor cells. These researchers also found some things in the genome that had never been seen in a prokaryote before. One of the key features that separates eukaryotes (humans, plants, fungi, and so on) and prokaryotes (bacteria and archaea) is that we eukaryotes can move things around on little conveyor belts inside our cells with special proteins. Lokiarchaeota have genes that encode those proteins, which is another clue that they belong on our team. The researchers also found genes suggesting they might have passed hydrogen back and forth with an endosymbiont (maybe a proto-mitochondria), as well as external structures that would help them hang on to their Alphaproteobacterial friend.

To learn more about these promising creatures and their internal structures, scientists needed better images of them. But the only way to do this would be to grow Lokiarchaeota in a culture. Despite all the cultures people had made from hydrothermal vents, Lokiarchaeota had never been among them— that is, until 2020.

The effort to culture Lokiarchaeota started way back in 2006, nearly a decade before Spang and her colleagues announced the first genome of Lokiarchaeota. Hiroyuki Imachi and his colleagues at the JAMSTEC decided to try a new approach to growing them. They built a special closet-sized chamber threaded with little pieces of sterilized kitchen sponges hanging along a nylon string. They soaked these sponges in fluids from a deep-sea hydrothermal vent off the coast of Japan. Then they dripped growth media slowly onto the top sponge. Once the first sponge was saturated, the liquid dripped excruciatingly slowly down to the next sponge and so on. Then they waited, patiently. Fourteen years later, Hiroyuki and his coauthors reported the first culture of Lokiarchaeota,[4] which they were able to photograph in high resolution.

The results delivered the kind of storybook ending that scientists dream about: it turns out that there are large appendages protruding out of each little Lokiarchaeota cell. If you use your imagination, you can just imagine these cells using their armlike thingies to tangle up an Alphaproteobacterial cell, engulf it, and eventually evolve into a giraffe. I watched over Zoom as Hiroyuki described his group's work while he received an award from the International Center for Deep Life Investigations in China for the discovery. He mentioned that the composition of the culture media had been partially inspired by some of the research I had done previously, and I was proud to have contributed, even if I did so unwittingly, to his team's success. They named the new organism *Prometheoarchaeum syntrophicum*, after Prometheus, the Greek god who forged humans out of mud and gave them fire (a fitting name if we guessed right about it being the progenitor of all eukaryotic life and giving us the firepower of mitochondria). The fact that we keep finding Lokiarchaeota over and over again in hydrothermal vents seems like a potential indicator that deep branches of life evolved in this environment. But we certainly can't go back a billion years and test the hypothesis. In origin of life research, we have to learn to live with a little ambiguity, especially when it comes to tackling the more complicated questions. In this chapter, I've gone through some plausible scenarios for where and how life began. Now it's time to move on to those more complicated questions: *When* and *why* did life originate?

10

EQUILIBRIUM IS DEATH

AS I'LL describe in this chapter, we can use chemical fossils to put a time stamp on the earliest life on Earth, allowing us to start learning about *when* life formed. Finding out *why* life formed is much harder. For this, we have to rely more on theory than data, and theory leads us to the shocking conclusion that life may be on a *continuum* with nonlife, rather than being two separate things. Let's dig in.

When Did Life Originate?

When did our world transform from a place where *nothing* was alive to a world where *something* was alive? That must have been one hell of a day. Finding fossils from the first life on Earth would be a convenient way to learn about when this happened. It would be nice if we could carefully chip into a rock and see physical evidence of life's origin as easily as we can exhume a *Tyrannosaurus rex* skeleton. But *T. rex* showed up relatively recently, less than 100 million years ago. So, *T. rex* fossils are still available to be found. In contrast, we're hunting for something from *billions* of years ago, and that thing, notably, lacked bones. We do have some ancient fossils that look like bacteria, called microfossils, but it's hard to distinguish their shapes from mineral precipitates. To hunt for the origin of life, we pair these microfossils with *chemical* fossils.

One of the most powerful chemical fossils available in the search for life's origins is something called elemental isotopes. Elemental isotopes are different atoms of the same element that differ only in the number of neutrons in their nuclei. If you look at a periodic table, you can squint and see a tiny little number under each element. That's the sum of neutrons and protons that are in that element's nucleus.* But a weird thing is that the number is almost always a non-integer. Carbon, for instance, has an atomic mass of 12.011. Every carbon atom must have 6 protons; otherwise it wouldn't be carbon. But that means that to add up to 12.011, carbon would have to have 6.011 neutrons, and this is impossible. There's no such thing as 0.011 of a neutron. What's going on here? The number 12.011 is not, in fact, the weight of any specific carbon atom, but rather the average weight of all the carbon atoms on Earth. Most carbon atoms have 6 neutrons, but a few of them have 7, and even fewer have 8 neutrons. As a result, the sum of protons and neutrons usually varies from 12 to 14, with an average of about 12.011.

Life is a master at detecting these subtle differences in carbon's neutron content. And when life interferes with chemistry, it makes an imprint we can see—as in a fossil. Carbon atoms with 8 neutrons are a special radioactive case, meaning that they decay spontaneously to other elements while releasing a blast of energy, but radiation is not what we're concerned with for chemical fossils. When we're using isotopes to detect life, we care about the fact that having more neutrons makes an atom

---

* Technically that little number is total atomic mass, which also includes electrons, but we can ignore them because their weight is inconsequential relative to that of neutrons and protons.

bond more tightly with other atoms. This minuscule difference in the bond strength of isotopes is inconsequential for the atom's chemical properties, but it is the secret to using these isotopes for life detection.

How exactly do we use isotopes as chemical fossils? As we know, life takes carbon dioxide and turns it into all the organic molecules it needs (sugars, lipids, proteins, nucleic acids) through autotrophy (either chemolithoautotrophic or photoautotrophic). And when life makes a random grab of carbon dioxide molecules, it mostly catches a mixture of 6-neutron and 7-neutron carbon atoms. Both isotopes work just fine, but the weaker-bonding 6-neutron versions react faster, leaving the microbe with a higher content of 6-neutron carbon in its body than the original mixture. Therefore, when we discover organic carbon in an ancient rock that has, say, an average neutron + proton number of 12.005 instead of 12.011, then we know that something living must have caused this shift.

But to put dates on these chemical fossils, we have to know how old the rocks are that harbor them. And how do you date a rock? The answer, again, is isotopes, this time of the radioactive kind. Radioactive isotopes (or radioisotopes, for short) are unstable, meaning they decay with a regular frequency over time. Usually when we think of radioactive decay, we're mostly concerned with the energy that shoots off from each decay event, because it could either give us cancer (e.g., the Chernobyl fallout), cure cancer (e.g., radiation therapy), or make food for microbes (see chapter 1). But to figure out how old rocks are, people use types of radioactive decay that are way, way slower than the ones that zap cancer. By measuring how many decay events the radioactive isotopes in a rock have experienced and knowing the rate of decay of that isotope, scientists can determine how old the rock is. Luckily, different kinds of rocks contain radioactive elements that decay at different rates, so by using different isotopes, we can cover pretty much the full range of timescales on Earth.

When a rock forms, say, at the seafloor, as the mantle bursts up between two tectonic plates that are pulling away from each other, the radioisotopes are reset, and the countdown begins. This means that there are little isotopic clocks embedded throughout Earth's crust, running at different rates, that tell us how old each rock is. In young rocks, we can determine their ages by measuring the fast-decaying radioisotopes, such as carbon. In extremely old rocks, the slowest-decaying elements, such as zircon, must be used. The accuracy of the measurement suffers in the slow-decaying radioisotopes, but these estimates can often be confirmed with other bits of information, as rocks are almost always embedded within formations of other layers of geological material. If your radioisotope measurements tell you that a rock is 3 billion years old, but it's wedged between coal seams, you know your measurements are wrong because coal comes from dead plant matter and plants only evolved ~150 million years ago. But if the isotopic evidence tells you that a rock is old and it's in the vicinity of other rocks that are also plausibly old, then you have a great find on your hands.

The oldest isotopic fossils on Earth are about 3.85 billion years old. Most other questions we have about the origin of life, such as where it started, have very squishy answers, but the date for the first evidence of life on Earth is solid. And it's an impressive number. Animals only appeared a few hundred million years ago. This means that for billions of years after life evolved, the only life on Earth was microbial. So when we study intraterrestrials and other microbes, we're studying lifestyles that have always dominated this planet, even though we only discovered them relatively recently.

But Earth was a planet for almost 700 million years before our oldest isotopic fossils appeared. What was happening then? Even though Earth is about 4.5 billion years old, the reason our rocks only go back to ~3.8 billion years old is because Earth's surface is constantly being reworked by plate tectonics. New rocks are formed at mid-ocean ridges while old ones are destroyed at subduction zones as tectonic plates are buried into the mantle. However, through happenstance, some small pieces of Earth's crust haven't yet been banished to oblivion in the mantle. It's as if you put together all the ingredients of cake batter and stirred really slowly. Your spoon could make quite a few trips around the bowl and still leave a bit of intact egg yolk not yet folded into the batter. It's the same with ancient rocks at Earth's surface. Some of them have escaped the swipe of the subduction spoon. Lucky us! In a few rare places—South Africa, Australia, and Greenland—we can find bits of rock that have avoided the fates that befell all the other rocks. And nearly every time a scientist looks at one of these rocks, they find little structures that look like they could be fossilized microbes, with isotope signals that look like life too. So it seems as though Earth had life as far back as its oldest rocks. Anything older than that lives only in our imaginations.

Why Did Life Form on Earth?

To figure out *why* life formed on Earth, let's go back to thermodynamics, since it's the powerful boundary-delineator for life. However, if we're going to use thermodynamics to try and find out the *reason* for life, we're going to have to push well past what J. Willard Gibbs dreamed up.

One person who had an especially wild idea for the origin of life was Vladimir Vernadsky, who boldly asserted that *there is no origin of life*. Period. In his view, life is on Earth because life has *always* been on Earth. He believed that the universe is eternal, so life must be too. Of course, Vernadsky was writing when the timing of Earth's formation was not well-known. Now we know that Earth has an origin, so life must have one too. But maybe there's something to be gained from thinking about this non-origin origin of life. Maybe we shouldn't be thinking of life as some special, highly rare event that happened once in this little section of the universe. Maybe life is more of a continuous series of events.

Another scientist, Alfred Lotka (1880–1949), is primarily known for developing the Lotka-Volterra models that show that populations of predators and prey oscillate relative to each other. But he also suggested that life was more of a "process" than a singular incident because it's constantly being replenished by energy coming from the sun.[1] I think that he, like Vernadsky, was on to something.

To understand the processes that Lotka was talking about, we need to look at some glaring problems with the Gibbs Free Energy equation I presented earlier. Gibbs Free Energy may be the most essential rule for life, but it is terrible at *describing* life. ΔG tells you how much energy you'll get when you run a chemical reaction all the way to equilibrium. But equilibrium is death. Gibbs Energy would do a fine job of describing my life if I were sealed in a gas-tight box. I would breathe up all the air in the box, ΔG would run to zero, I would die, and the second law of thermodynamics would be fulfilled.

But life does not happen in a box. Life is an open system. Energy is constantly flowing through our bodies, sustaining us, and pushing us back out of equilibrium. Trying to divine the thermodynamics of life from Gibbs Free Energy alone is like working only with museum specimens. From a taxidermized bird, insect, and hippopotamus, we might guess that a bird eats with its beak, an insect's exoskeleton gives it structure, and a hippopotamus listens with its ears. But how would we ever know about the satisfying crunch the bird experiences as it picks insects off the hippo's ear? Equilibrium thermodynamics is powerful and useful, but it's not the whole shebang.

To describe open systems that are far away from equilibrium (that is, the ones that are alive), we need an entirely different type of thermodynamics. We need something called non-equilibrium thermodynamics. In their 2005 book, *Into the Cool*, Eric Schneider and Dorian Sagan put it nicely: "Evolutionary theory links organisms in time. Ecology links organisms in space. Chemistry links them in structure. Non-equilibrium thermodynamics links them in process." Non-equilibrium thermodynamics, they argue, is as important to life as chemistry itself.

To understand why non-equilibrium thermodynamics is so important, consider the significance of the DNA revolution. This breakthrough was powerful because it showed us beyond a shadow of a doubt that humans are evolutionarily related to all life, including every single microbe. You can line up DNA sequences from every living organism on Earth and clearly see that we're all kin. Non-equilibrium thermodynamics extends this kinship to include the *nonliving world itself*. Schneider and Sagan suggest that life and nonlife exist in a continuous line. At one end of the spectrum are nonliving systems at thermodynamic equilibrium, and at the other end are living systems that are not only out of equilibrium but are continuously creating further energetic potential to make sure they stay well out of equilibrium. Where things sit on that continuum depends on how each of them works within the boundaries of non-equilibrium thermodynamics. In this way, life and nonlife are fundamentally connected by the second law of thermodynamics, so if we want to use science to find the "why" of life, we have to start with this law.

The second law of thermodynamics states that a nebulous thing called entropy must always increase. Entropy, according to the *New Oxford American Dictionary*, is "a thermodynamic quantity representing the unavailability of a system's thermal energy for conversion into mechanical work." If you multiply entropy by temperature, you get energy, so entropy is not energy— it's more like the *quality* of a given amount of energy—and when entropy is very high, the energy is less useful, like heat or water we lose during a workout. Entropy can be viewed as possibility, and systems must go toward what is more possible. This is why the second law of thermodynamics says entropy has to increase. The dictionary's definition of entropy continues: "often interpreted as the degree of disorder or randomness in the system." If you drop a bowl of soup, a disordered high-entropy mess is going to splash all over your shoes. Disorder is such an intuitive way to understand entropy that, unfortunately, it has come to define it in many people's minds. However, to view entropy solely as a measure of disorder or chaos would be a grave mistake. Here's an illustration to show why disorder is an intuitive way to understand some types of entropy yet fails to describe entropy in all situations.

If a helium balloon pops, the helium molecules will rapidly spread out to fill the whole room, reaching a maximum state of disorder. Here, disorder can be a stand-in for entropy because they both increase together. But that's not the whole story. About the same time that J. Willard Gibbs was busy deriving thermodynamic equations, Ludwig Boltzmann came to the uncomfortable conclusion that if you want to determine the collective behavior of all the helium molecules in the above example, you can't add up the individual behaviors of each molecule like they're little balls flying around in the air. Instead, to make the numbers come out correctly, you need to apply the laws of statistics and probability.

This basic idea—that we must incorporate chance into physics—was wildly unpopular among many physicists at the time since it felt like admitting failure to control every variable of the physical world. But Boltzmann would ultimately prove to be correct, starting us on the path to modern quantum physics, a science based almost entirely on probabilities.

One thing that allowed Boltzmann to make this breakthrough was his fascination with Charles Darwin's ideas about biological evolution. He was inspired by Darwin's account of how natural selection works on large populations of animals, and Boltzmann applied this thinking to the ways in which thermodynamics works on large populations of molecules. Boltzmann once said, "If someone asked me what name we should give to this century, I would answer without hesitation that this is the century of Darwin."[2] I can't think of a better illustration of the lack of barriers between the physical and biological worlds than the fact that the work that paved the way to quantum physics was based on evolutionary biology.

To understand how probability is involved in thermodynamics, let's go back to the helium balloon example. It makes probabilistic sense that disorder must increase in lockstep with entropy because there are many more ways that molecules of helium can be distributed all around the room compared to the number of ways they can be distributed in a balloon-sized pocket of the room. The molecules spread out because there are more options for spreading out than there are for remaining in one tiny place.

In this way, the second law of thermodynamics is not mystical—it's obvious. This "gas-in-a-box" example is so comfortingly intuitive that it can be tempting to assume that all processes increase entropy by increasing disorder. But this is simply not the case. In *open* systems, for instance, entropy can come from a *decrease* in disorder if it creates more entropy outside the system, for instance, by releasing heat.

Let's return to the analogy of a helium balloon popping in a room. This time, however, let's imagine that room also contains a ten-year-old child who sucks the helium out of the balloon. In that case, instead of the gas molecules spreading out to the whole room, they would instead go into a smaller, more ordered space in her lungs, so she can talk with a funny high-pitched voice. Has the child broken the second law of thermodynamics by preventing the gas molecules from filling up the room when the balloon is opened? No. When she dissolves into giggles because she sounds like a little chipmunk, she makes heat. Even though she delays the increase in disorder of the helium molecules by pulling them into her lungs, more total entropy is produced by her laughter when she finally releases them to disperse.

Gas-in-a-box is not a generalizable example, but gas-in-a-box-plus-giggling-child may have some legs. It might help us understand how life is a natural outcome of the second law of thermodynamics: life is extremely good at continuously creating structures that make more opportunities for entropy production. Now, what about something that's clearly not alive? Does nonlife have similar entropy-creating processes that are also ordered? The answer is yes. An eddy in a stream is not alive; however, it has a highly ordered structure. Water molecules usually move separately, but for a short amount of time they become a highly ordered whirlpool to maximize entropy production through energy dissipation from the system. The water molecules on the inside and outside of the eddy can't talk to each other or make decisions, but they retain the same angular momentum to keep the structure intact for a brief period of time.

Ilya Prigogine (1917–2003) named phenomena like whirlpools "dissipative structures" since they exist to dissipate energy and they maintain a surprisingly low-entropy, ordered state. But dissipative structures only seem surprising if you're missing the larger picture of the open system in which they occur. Prigogine won the Nobel Prize for showing that when you're far from equilibrium, entropy production creates structure and order, not disorder, which is an abiotic mimic of life. He conducted his work in autocatalytic* systems where two competing chemical reactions, each associated with a color change, sync up and cause colorful structures to form in lifeless liquids. This was, he showed, an extreme case of an "eddy in a stream" where the nature of the oscillating reactions demands that they turn colors at regular intervals, making chemical clocks. In an open, autocatalytic system, with many possible paths, structure not only *can* arise, but *must* arise—we just don't know exactly what form each structure will take.

A wide variety of objects (living or not) produce order as an emergent property. A crystal, for instance, is a highly ordered structure with a regular repeating pattern like wallpaper. Living cells can also be viewed as highly ordered, with lipids organized into cell walls, DNA tightly wound, and proteins compartmentalized into the places where they're needed. Each organ in the human body functions only because it is organized in a nonrandom way. Almost any phenomenon of life can be understood as a dissipative structure. Wars and trade agreements are both

---

* Autocatalytic means that something helps create its own self. People are autocatalytic because we create our own babies. Chemical reactions are autocatalytic when the product interacts with reactants to make more product.

structures that dissipate energy, although the first brings death and chaos and the second brings organization and stability. The arrangement of plants in a forest is a dissipative structure since plants organize the placement of their leaves to capture more photons of light, thereby creating further entropy throughout the plants' lives and afterlives as they become food for other organisms. The molecules inside the plants' cells that are used to capture this light are also ordered, but the molecules themselves are nonliving material. No one would ever say that chlorophyll itself is alive.

If life and nonlife are on a continuum, how can we scientifically distinguish between the two? According to non-equilibrium thermodynamics, it is life's propensity for continually pushing systems back out of equilibrium that sets it apart from nonlife. Both life and nonlife create ordered structures to maximize entropy production within a system, but only life *creates new systems* in which entropy can be produced. Nonlife makes whirlpools, but life dams the river to make more whirlpools so it can go whitewater rafting on them.

How Do Aeonophiles Fulfill the Second Law of Thermodynamics?

If life exists to create new opportunities for entropy production, is it possible that these ultradormant intraterrestrials— the aeonophiles—have found a way to create entropy-producing non-equilibrium systems that we've never seen before?

Joseph Vallino, a scientist at the Marine Biological Laboratory in Woods Hole, Massachusetts, suggests that life spreads out entropy production over increasingly longer timescales to maximize total entropy production over the long run.[3] It is, he once said to me, kind of like the difference between getting a job after high school versus going to college and getting a job afterward. You make more money in the short term by skipping college and going to work, but over longer timescales, on average, you make more money in jobs that require a college degree.* Abiotic systems get a job right out of high school, while biology goes to college, using money as a stand-in for entropy. In *Into the Cool*, Schneider and Sagan make a similar point. As they note, the second law of thermodynamics doesn't say how fast entropy has to be produced. Life can spread it out a bit, delaying the progress toward equilibrium, milking it for all its worth, "trapping and rerouting" it to make more life and to create more total entropy in the process.

This means that the point of life may be to spread phenomena out over increasingly longer timescales. You know who likes long timescales? Aeonophiles! Humans increase the timescales over which we pass information to each other by writing books that are read across many generations. Religion, in particular, seems to have been a powerful innovation for making people feel intimately connected to people that lived thousands of years ago. Intraterrestrials introduce a fundamentally different way to spread out entropy production over longer timescales. Intraterrestrials do this simply by living long enough to span a long timescale as an individual. This is a different way to push the system back out of equilibrium than the ones we encounter in our daily lives. Aeonophiles spread out the entropy produced through metabolic heat loss over timescales too long for a human to relate to.

---

* There are, of course, many people who have made more money without a college degree. This is where Boltzmann's probabilities come in—on average, people with college degrees make more money, even though some individuals break the trend.

Even though they may seem extraordinary to us, individuals that live for thousands of years or longer may be ordinary on Earth. In fact, they might be key to understanding *why* life exists. Finding such an outlandish new way to create opportunities for entropy production over very long timescales supports the idea that this opportunity for entropy creation is, itself, the *why* of life: Life happens because the second law of thermodynamics demands it.

II

WHAT CAN INTRATERRESTRIALS DO FOR US?

BURIED IN the deep dark subsurface, with their own strange sources of food and energy, producing entropy over timescales we can't relate to, intraterrestrials have remained largely remote and aloof from us humans. As a result, it may seem like they would have little opportunity to directly impact our own lives.

On the contrary, the microbes that dwell in the deep subsurface biosphere affect our lives in innumerable ways. On a planetary scale, for instance, they are responsible for the oxygen that we breathe. Most of our oxygen is produced by plants and plankton, but it would just recombine with plant matter and turn back into carbon dioxide and water if it weren't for subsurface microbes that sequester the organic matter that would otherwise react with it. In addition, without the nutrients recycled by intraterrestrials in the seafloor, such as iron and nitrogen, phytoplankton would be severely limited in their ability to make oxygen for us. Finally, intraterrestrials are uniquely suited to detoxify our worst wastes—by breathing radioactive uranium, arsenic, organic carcinogens, and other nasty stuff—which may allow us to develop as a society without killing ourselves in the process.

It's not all roses and sunshine, of course. We are locked in a constant battle with the seafloor intraterrestrials that dissolve ship hulls, oil rigs, and bridge supports with glee. They're also great at making methane in the subsurface, which can explode during mining and oil drilling, in addition to being a greenhouse gas.

Intraterrestrials also hold great power when it comes to climate change. Their relationship to climate change may be the most important connection they have to us surface dwellers. In this chapter, I'll describe my ongoing work in the high Arctic permafrost, where we are attempting to determine whether intraterrestrials are turning the carbon from thawing permafrost into greenhouse gases that might end up exacerbating climate change. Next, I'll reveal how intraterrestrials may affect the carbon storage technologies that are being developed to mitigate the effects of climate change. Finally, I will talk about a pressing issue today: deep-sea mining—the mining of precious metals to drive renewable energy, which may affect the health of the oceans. This type of mining may ultimately destroy intraterrestrial ecosystems before we even have a chance to discover what they're capable of.

Ny-Ålesund, Svalbard, 79°N

One thing I love about fieldwork is that no matter how many fancy academic degrees you have, no matter how smart you think you are, nature will find a way to lay you low. It's humbling. I was reminded of this in March 2021, when my colleagues and I were drilling in permafrost, which is soil that never thaws, even in the summer. It exists in the Arctic, the Antarctic, and high-altitude locations, and it lies underneath upper "active-layer" soils that thaw every summer. I'd worked on permafrost samples in the past, ones that other people had brought to me, but I had never actually drilled into the stuff myself. And as far as I knew, no one else had yet looked for intraterrestrials in the permafrost in Ny-Ålesund, Svalbard. So, another adventure was in order.

In the summer of 2019, when the upper active layers were thawed, I joined Tatiana Vishnivetskaya from the University of Tennessee and Andrey Abramov from the Institute of Physico-chemical and Biological Problems in Soil Science in Russia to do some reconnaissance. Permafrost thaw due to climate change is more complicated than simply turning frozen ground to mud. The soil above permafrost thaws every summer, which is a natural phenomenon. The real permafrost lies deep beneath these active-layer soils, and this part stays frozen year-round, so the microbial activity is low. The question is, when permafrost thaws from climate change, how much greenhouse gas will be emitted from microbial activity? In Ny-Ålesund, the active layer is quite thick; about two meters deep. The permafrost that lies below it has all the makings of an intraterrestrial paradise—no sun, no fast-moving creatures to compete with, and not much to eat because the food is frozen solid. Unfortunately for us, it is also well fortified against meddlesome scientists. Permafrost, unlike a subduction zone, doesn't have the lucky happenstance of being dotted with springs that naturally flush intraterrestrials to the surface. To sample permafrost, we have no choice but to drill down into it.

Tatiana, Andrey, and I were there in the summer to try to anticipate any problems, such as the presence of rocks, that would arise when we did the real drilling under snow cover in the winter. And indeed what we found were rocks—lots of them—enough to potentially stop even our powerful drill. So that summer, we poked around for the most rock-free zone and marked the location. When we returned in March 2021, after a long COVID-19 quarantine in Oslo, we were ready to drill.

Fortunately, my colleague Andrey is not only an expert in permafrost, he's also an expert in drilling. The drill he uses is a Soviet-era rig, and it works well. Its four-stroke Briggs & Stratton engine spins a drill head attached to a core barrel lined with jagged metal teeth. It also has a series of pulleys that put downward or upward pressure on the core barrel while it rotates. If all goes well, you can screw on as many drilling rods as you want in a series, which allows you to push the core barrel ever deeper into the permafrost. But as you might expect, nothing is ever easy in the field.

In addition to the rock problem, there is the issue of the cold temperatures. When the barrel heats up from the friction of the drilling, it can thaw the water in the soil. If you don't keep the barrel in motion, this water will freeze the core barrel into the ground like a popsicle stick. So why don't we just wait until summer when it's warmer, and attack the thing with shovels? For one, even though it's warmer at the surface, the deep permafrost is still frozen, so at some depth, the same problem would arise. But the bigger reason is that dragging a heavy drill across fragile Arctic plants is a nonstarter. Once the protective blanket of snow arrives in the winter, scientists can transport their equipment with a snowmobile and a sled.

In 2021, after I finally got the hang of driving a snowmobile, we pulled the massive drill, core barrels, and drilling rods out to the drilling site on sleds. We were completely surrounded by white. Crisp white snow shushed under our snowmobile tread, brilliant white peaks shone off in the distance, white glaciers blanketed the valleys, and white icebergs dotted the dark blue fjord. It was stunning. Luckily there were no hungry polar bears nearby—they would have been invisible against the landscape.

The drilling itself seemed to be going well at first. While the rocks put up a fight, Andrey's masterful drilling techniques dragged up core after core, often boring straight through big rocks with brute force. The sun was shining, the drill wasn't overpressured, the cores weren't freezing to the ground, and I had finally mastered using the unwieldy giant pipe wrench to unscrew the drill bit and pull out the core liner of each precious section as it came up. It was the perfect setting for a theatrical downfall.

The core barrel and the drill rods that we add to it when the hole gets deeper are each topped by a metal screw that goes into the drill head. As we were happily drilling away that day, without warning, the screw broke, disconnecting it from the drill rod and leaving half of the screw threaded deep within the drill head.

Luckily, we were able to coax the broken screw out by tapping on it from the inside of the screw fragment with a frozen screwdriver and a frozen mallet. We then spent the rest of the day trying to determine what had gone wrong. It looked like the metal itself had failed, even without much torque. The extreme cold may have been more than the metal could bear. We were spending whole days out in the −28°C weather, and the cold may have weakened the metal. Normally, metal drill bits can handle extreme cold, but maybe we ended up with lower quality alloys. There was nothing we could do but choke down bites of our frozen sandwiches and try again.

We tried another core barrel. It retrieved one core and then broke. We tried another core barrel. That one broke. Then another. And another. We were running out of core barrels. We managed to identify one drill rod that was capable of retrieving cores without breaking. But we still needed a core barrel with a functional screw. This is when my PhD student at the time,

Katie Sipes, radioed in from the laboratory to say, "I don't know if this is helpful, but I know how to weld." Welding, it turns out, was the answer. Now we had a working rig.

We started the next day with high hopes at our second location, Kvadehuken, a starkly beautiful peninsula that is about a thirty-minute snowmobile drive from Ny-Ålesund. Here, we managed to drill down to bedrock past the all-important two-meters depth of the start of the permafrost. Success! I did a full-on happy dance when we hit two meters. That was the goal of the trip, and we had achieved it. But we were drilling into a shallow brine lake, so the cores were coming up wet, which is fairly dissatisfying when one is going for the frozen stuff. I am told that it still technically qualifies as permafrost because it is below the freezing point of water. Whatever. I don't make the rules.

To feel satisfied, however, we needed something with rock-hard ice in it. Something that looked like the permafrost of our dreams. For that, we went back to a site near the Amundsen-Nobile Climate Change Tower run by the National Research Council of Italy. Here, a core barrel and its drilling rods refrained from breaking long enough to hit the jackpot: an ice layer that glinted like diamonds, with solid frozen soil beneath. The ice line was right at two meters, where previous data predicted it would be. Despite all our tough breaks, through the ingenuity and good spirits of our team and the amazing research support in Ny-Ålesund, we still managed to bring back a freezer full of samples.

But is anything alive in permafrost? The process of freezing creates little ice crystals that kill bacteria by skewering them—we depend on this antimicrobial process when we store food in our freezers. However, maybe permafrost microbes are a little different compared to the ones that can spoil a ham hock. A bigger problem than little ice daggers is that life requires liquid water to function—no exceptions, even for the intraterrestrials. It may therefore seem like permafrost would be a death trap for microbes, but the important thing to know about permafrost is that, even when it is frozen solid, it still contains minuscule amounts of liquid water. Salty water freezes at much lower temperatures than pure water. That's why we throw salt on sidewalks before a snowstorm. When soils freeze, the pure water crystallizes into ice, leaving behind any salts that were present. This salty leftover water can stay liquid all winter long, or however long the permafrost stays frozen. And, on Earth at least, whenever we find liquid water, it almost always has something living in it.

Currently, all signs point to the existence of living microbes in permafrost, even in some of the oldest permafrost in the world, which is in Siberia, where soils have been frozen solid for at least 1.1 million years. Their DNA is intact, not broken to bits like decayed body parts. And this DNA is inside cells, which must be intact too; otherwise the DNA would have leaked out.[1] The cells even seem to have a proton motive force, which, as I explained in previous chapters, is a quintessential feature of life as we know it. The nature of the microbes themselves argue against the possibility that they just fell in there and died; they have genes adapted to survive in these thin brine veins in permafrost.[2] I am in awe of a microbe that can live in a tiny pocket of salty water for hundreds to millions of years— they may be additional members of the aeonophile club.

As is true for many creatures on our planet, the habitat within which these microbes live is changing rapidly. Julia Boike, a scientist at the Alfred Wegener Institute in Germany, established the permafrost monitoring site called Bayelva near Ny-Ålesund in 1998, which has produced one of the longest-running datasets in the world for permafrost snow height, temperature, and moisture content. Her results, like those of other scientists around the world, show that permafrost is warming and thawing from climate change. These changes, driven by the increase in greenhouse gasses in our atmosphere, would be much worse were it not for phytoplankton in the oceans, working overtime to sop up our mess.* Eventually, this excess carbon will trickle down to the subsurface and the chemolithoautotrophs will pitch in to the effort. But that could be millions or hundreds of millions of years from now. Permafrost, on the other hand, is shifting right under our noses.

Much of the carbon on Earth's surface is bound up in permafrost, so when it thaws, subsurface microbes might increase their metabolic activity, which would convert this soil carbon to methane and carbon dioxide, both of which are molecules that can exacerbate climate change. Almost every summer, new patches of previously frozen ground thaw, so new layers of microbes are waking up every year. And we don't yet know enough about them to know what they're going to do when that occurs. Currently, about 1,600 petagrams of carbon are sequestered in permanently frozen soils in Siberia, Greenland, Canada, Alaska, Antarctica, high-altitude locations at lower latitudes, and even underwater.[3] This number is about twice the amount of carbon currently in the atmosphere, so turning it all into gas would be catastrophic.[4]

Luckily, things aren't that simple. The consequences of the thawing permafrost are exceedingly difficult to predict, as is the case with the countless other effects of climate change. The possibilities range from the very bad (all the carbon in the soil

---

* Of course, they are not helping us on purpose. But without the algae and cyanobacteria in the world's oceans, carbon dioxide concentrations would be MUCH higher than they are now.

changes to methane, driving a massive amount of further warming), to neutral (some methane and carbon dioxide are produced, but emissions are kept under control by other microbes that use them up), to helpful (thawing permafrost starts drawing down atmospheric carbon dioxide levels because it's full of chemolithoautotrophs or it stimulates plant growth). A runaway positive feedback loop between permafrost and the atmosphere may not be our fate. Most likely, there's not a one-size-fits-all answer; the consequences will differ depending on latitudes, altitudes, rainfall amounts, and yearly temperature changes. And we're not going to find the answer by thinking really hard about the problem, or by relying exclusively on experiments in laboratories. We need to continue to go out into the natural world to see what's really happening out there.

Currently, most global climate change scenario models do not even consider the influence of permafrost intraterrestrials. It's not that the modelers haven't heard of permafrost microbes or that they don't think they're important. It's that global climate models are not yet able to handle this level of nuance. When I'm out in the Artic, getting blasted with −50°C gusts of wind, cursing rocks and flimsy drill rods, I'm building knowledge—not just about how intraterrestrials withstand these harsh conditions, but about whether they are likely to take some of the sting out of climate change or, in the worst-case scenario, spell our doom. When the models are ready for us, we need to be ready for them.

Permafrost microbes aren't the only intraterrestrials that have a role to play in our current climate crisis. Microbes in deep aquifers and tiny rock fractures may serve as sources of salvation. This should be unsurprising: most of the excess greenhouse gasses currently in our atmosphere today came from the burning of fossil fuels that came from underground.

One obvious solution, then, is to find a way to push those nasty gasses back from whence they came through a process called carbon capture and storage (CCS).

The main snag with CCS is that the gasses were unearthed in the form of petroleum and methane, but we are returning them as carbon dioxide. Fortunately, there are plenty of types of rock that will accept large amounts of carbon dioxide, and certain countries, such as Iceland, are making significant headway in an area that once seemed unlikely.[5]

For CCS to be successful, this carbon must be stored deep within Earth's crust for thousands of years or longer, staying safely out of reach of the atmosphere. And if the subsurface were like an inert Tupperware container, then we could just pump that carbon down there, sit back, and watch the global temperature stabilize and the glaciers creep back to where they're supposed to be. But Earth's crust is not an empty vessel. In addition to all the chemicals in the rocks that might interact with the injected carbon dioxide, mercurial intraterrestrials are also hiding in every crack and crevice. Subsurface microbes might suck up this extra carbon from CCS and help turn it into solid rock, making sure it stays well away from our atmosphere for a very long time. Or they might turn it into rock too quickly as it's being pumped downward, gumming up the works. Or they might use hydrogen generated from rocks and combine it with carbon dioxide to make methane that could escape and accelerate global warming, since methane is a more powerful greenhouse gas than carbon dioxide. At this moment, we simply don't know what the intraterrestrials are going to do with the carbon we pump into the Earth to try and mitigate climate change.

Some early indications suggest that the intraterrestrials might not be our friends in this adventure. Certain CCS test wells confirm that they are starting to produce some methane.[6]

Other studies, however, show that natural intraterrestrials help precipitate carbon dioxide as carbonate rocks or consume it, much like forests do at the surface.[7] These microbes might happily chomp all this carbon up for us. If so, first we need to figure out the best way to capture this carbon and deliver it to them, and then we need to figure out which environments have the right kind of intraterrestrials to make the whole enterprise successful. We are just beginning to understand what the intraterrestrials are going to do with these subsurface storage tanks, but I'm hopeful that CCS will eventually help us slow the effects of climate change.

Deep-Sea Mining

In addition to putting carbon back into the ground, we also need to stop digging more of it up. We can capture energy from the sun, wind, tides, temperature gradients, dams, plants, and algae. However, for renewable forms of energy to replace fossil fuels, their transport and storage must be flexible; flexibility requires good batteries, and good batteries need metals. As I've mentioned, intraterrestrials have an intimate relationship with a wide variety of metals, many of which are commercially relevant. When intraterrestrials breathe these metals, their respiration often changes the metals' mineral forms and determines whether the metals stay sequestered in soil and rock or whether they leach out of it. In addition to using metals for respiration, intraterrestrials also need a range of metals to make their enzymes work properly. Enzymes are like the engines of any cell, whether it's microbial or human, performing whatever cellular function needs to be done, such as eating, breathing, growing, making toxins, or swimming. Because we humans live at the oxygenated surface, our enzymes mostly need iron, magnesium, and a tiny bit of copper. But intraterrestrials' enzymes must be tuned to more redox states, and their enzymes require a much larger range of metals. They need iron, magnesium, and copper like we do, but they also need nickel, molybdenum, selenium, tungsten, neodymium, and so on.

Currently, all the metals we use for the batteries that power our personal electronics (as well as the large lithium-ion batteries used for transportation and high-powered magnets for windmills, for example) come from mining. And we have been employing microbes to help us mine metals since mining was invented. For instance, when copper sulfide ores are mined for copper, they are placed into large "biopiles" where microbes slowly but relentlessly pull the copper molecules out of their strong sulfide bonds and oxidize them. The microbes, of course, don't know that we want the copper. They're just trying to breathe, as transferring electrons from copper to oxygen has a huge and negative ΔG. In the process, the copper is released into water, which can be collected and further refined to make things like plumbing pipes and electronics. Microbes' abilities to concentrate minerals into useful stockpiles may determine whether our mineral resources on Earth are renewable or not. Unfortunately, the presence of microbes can also exacerbate the pollution from our mining operations. After the commercially viable metals have been leached out, mines are stuck with leftovers rich in iron and sulfur called tailings. These are like a smorgasbord for microbes that turn the whole mess into acid, contaminating nearby streams and groundwater, killing wildlife and polluting drinking water.

Most oceanographers know that metals aren't just found in mountains or ore seams in the Australian outback. A lot of the metals are in grapefruit-sized nodules at the bottom of the ocean and in metal-rich precipitates at hydrothermal vents. One of the first scientific oceanic voyages, called the Challenger expedition, pulled up several black spherical rocks that were sitting at the seafloor. These rocks, called manganese nodules or polymetallic nodules, form when something, a clam shell or a shark's tooth, provides a surface where metals can precipitate, causing a chain reaction in which the new precipitates catalyze even more metal deposition out of the seawater. Over time, this process eventually builds a round polymetallic nodule containing high concentrations of manganese, iron, cobalt, copper, and nickel. This metal-plating process is extremely slow. It takes about a million years to create a layer of metals on the nodule that is as thick as a layer of nail polish,[8] so, just like fossil fuels, these polymetallic nodules are not a renewable resource; if you scooped them all off the seafloor, you'd have to wait longer than humans have been in existence to detect a noticeable replenishment.

For decades, researchers and entrepreneurs have probed these nodules to figure out whether anyone could make any money mining them. So far, the nodules have been safe from mining because working in the deep sea is expensive and difficult. First, there's the cost of stationing a giant boat at the surface. Second, there's the even greater cost of developing new technologies to retrieve the material from the seafloor and process it. And every time a storm rolls through or the seas get rough, you'd have to pull up all your profit-making equipment from the seafloor, close up shop, and burn fuel to run away from the weather. In the past few years, however, precious metals have become increasingly important, drastically changing the economic calculus around deep-sea mining, and the industry is rapidly growing.

Every country with a coastline has an exclusive economic zone (EEZ) extending 200 nautical miles (or, 230.16 miles) into the ocean, where they control mining and fishing. But 54 percent of the ocean is too far offshore to belong to any country. So technically someone could drop a deep-sea dredge over the side of a ship out in the middle of the ocean and take what they want, without running afoul of any government's laws. To address this, the United Nations Convention on the Law of the Sea established the International Seabed Authority (ISA) in 1982 to "organize and control all mineral-resources-related activities . . . for the benefit of humankind as a whole."[9] The ISA is composed of 168 signatory nations (which includes the European Union, but not the United States of America) and is headquartered in Kingston, Jamaica. For better or for worse, it has decided to provide leases to mining companies in an area of the Pacific Ocean called the Clarion-Clipperton Zone, which is where much of the exploration for deep-sea mining is currently happening.

I am not thrilled by the idea of developing a deep-sea mining industry. I love the deep sea because it sits at the sweet spot between mysteriousness and accessibility. It's one of the least understood places on the planet, and we're still exploring it and unraveling its secrets. Now companies and governmental agencies are dredging part of it to explore whether they can expand to start mining greater swaths of the ocean. They're not doing this for any increase of knowledge but for personal profit. You'd better believe I'm paying close attention.

Here's how polymetallic nodule mining will likely work if it goes into large-scale operations. A ship will lower a dredge the size of a couple of eighteen-wheel trucks onto the seafloor. The dredge will then move by itself along the seafloor, scooping up everything in the upper few inches of seafloor and funneling it into a pipe that shoots the material up to the ship. At the ship, nodules will be separated from sediments that will be dumped overboard. To turn a profit, a mining operation will need to crank through many tons of seafloor and pull out many tons of polymetallic nodules. That will produce enormous quantities of sediment "bycatch," If all that muck is sprayed onto the ocean surface, it will darken the upper ocean, killing off phytoplankton and harming marine animals. We need phytoplankton to be healthy because they sequester greenhouse gasses and make the oxygen and food that support the abundant life in our oceans, including all our fisheries.

There is no question that massive darkening of the upper oceans would be catastrophic for the planet and for us humans. For this reason, companies have proposed to pipe the sediments back down to the deep ocean. Delivering all the leftover mine tailings back down to the seafloor cuts into profit, but so would an ocean full of dead dolphins.

There is, however, currently no consensus on how deep the mining companies should go to release these sediments. Every extra inch will incur additional expense, so mining companies will be financially incentivized to dump it in water as shallow as they can get away with. Moreover, even if the sediments are returned to the very bottom of the ocean, the process will still create a silty plume. I've been to the bottom of the sea, and I can say with great certainty that it is chock-full of life. I assume these exotic animals would prefer not to choke on clouds of mud.

There are other environmental hazards associated with deep-sea mining. First, the removal of the nodules themselves can harm the deep-sea ecosystem. In a world ruled by water, sand, and mud, a hard surface like a nodule is a precious commodity for seafloor life to settle on. Second, a lot of important biological and chemical processes happen in the upper layers of the seafloor. Many of the nutrients that are eventually flushed up to the surface ocean—stimulating photosynthesis, fueling commercial fish populations, and regulating the gasses in our atmosphere—are released from the upper few layers of seafloor that will be destroyed during mining. Third, even if you find a good solution to the problem of piping mud back down to the seafloor, which may not be possible, the act of dredging itself will inevitably kick up a muddy haze that may spread out and affect a larger section of ocean.

Each of these environmental impacts is exacerbated by the immense scale of mining operations. To put this in perspective, the amount of seafloor currently contracted for nodule exploration is at least four times larger than all the seafloor impacted by offshore oil and gas, and at least twice as large as is currently dredged for bottom fisheries. The Clarion-Clipperton Zone, where companies are currently conducting "small-"scale mining operations, would stretch almost all the way across the lower forty-eight states of the USA. The entire footprint of current mining on land represents only about 3 percent of the area that may be covered by deep-sea mining if they scale their operations into a full-size industry. As a result, hazards are going to pile up, with significant negative environmental consequences for the deep sea.

Polymetallic nodules are not the only targets for commercial mining in the deep sea. Polymetallic sulfides, otherwise known as hydrothermal vents, and cobalt-rich ferromanganese crusts are in the crosshairs too. Hydrothermal vents form when hot, metal-laden deep subsurface fluids shoot like a firehose into cold, oxygenated seawater, which forces metals in the fluids to precipitate. Cobalt-rich ferromanganese crusts, by contrast, form in the areas of the seafloor where ocean currents have swept the area clean of any sediments. These currents expose basalt produced at mid-ocean spreading ridges, which can precipitate metals like cobalt, nickel, platinum, manganese, thallium, and tellurium out of seawater.[10] Both of these types of structures make good physical substrates for deep-sea mussels and clams, as well as provide metals that fuel microbial life in the ocean. Like polymetallic nodules, these ecosystems are quietly humming along at the seafloor, providing services that keep the whole ocean in proper balance. Seems like something we shouldn't just destroy willy-nilly.

These negative effects of deep-sea mining are compounded by the fact that the deep sea is so underexplored. Many of the geological, chemical, physical, and biological processes happening right now in the deep sea—processes that have been occurring for millions or even billions of years—have yet to be discovered. I know this is true because we find new things at the seafloor all the time. As I've described above, these newly discovered ecosystems hold clues to early life on Earth, but they also might carry genes or processes that are helpful for medicine or ameliorating climate change. But we'll never discover this good stuff if we wipe out the whole ecosystem first.

So the answer's simple, right? Just ban all deep-sea mining. Don't plunder our common resources. Unfortunately, it's not so simple. We need phones—we're well past the era of phones that are stuck to the wall, and we're not going back. More importantly, we need electric cars and wind turbines if we're ever going to rely more heavily on renewable fuels, and these technologies require metals.

Maybe we have other options. There are more metals in landfills than are currently in useful circulation. If we can get better about recycling our electronic waste, the pressure for mining will decrease. However, such recycling industries don't yet exist on the scale that we need them to. We need to develop these technologies or decrease our dependence on rare metals. For example, some iron phosphate batteries have already been prototyped. There may even be a solution in the subsurface itself. Large-scale storage of energy produced intermittently through renewable methods is one of the big drivers of battery production needs. Converting this energy to highly energy-dense hydrogen and pumping it underground is proving to be a potentially clean, inexpensive, and effective method for energy storage that doesn't require stripping the Earth of its mineral resources.[11] This is promising technology, and while it won't solve the lithium demands of producing new iPhones, it could decrease the metal demands of large-scale industrial storage of energy production from wind and solar energy. Of course, the intraterrestrials are likely to have as much to say about hydrogen storage as they do about carbon dioxide storage. Therefore, this emerging technology will need to involve researchers and technicians well schooled in subsurface intraterrestrial activities.

Deep-sea mining is moving forward—quickly—whether it's good for us or not. Beth Orcutt at the Bigelow Institute for Ocean Sciences, who is the director of the US National Science Foundation–funded Crustal Ocean Biosphere Research Accelerator, suggested in a recent conversation with me that consumer resistance is one plausible path for protecting these vulnerable ecosystems: "If society supports keeping non-sustainably sourced metals from the deep sea out of the supply chain, companies will hopefully not use them. Already, big companies like BMW, Google, and Samsung have signed a moratorium on accepting any metals from deep-sea mining unless sourcing becomes sustainable." Responsible mining certification, similar to existing certifications for preventing child labor, is also on the horizon and could incentivize companies to mine sustainably.

I bet that, given the nature of territorial rights over the ocean floor, some countries will decide to plunge into potentially hazardous mining. At the time of writing, Norway, Japan, and the Cook Islands have leased rights to their EEZ to mining contractors for resource exploration. Papua New Guinea originally supported mining their EEZ, but the company, called Nautilus,* pulled out when they determined it wasn't economically advantageous, leaving Papua New Guinea with a massive economic loss.

What, then, should proper international regulation involve? That has yet to be determined. Mining has a terrible track record in human history of consolidating profit for a few people while destroying the environment for everyone else. Deep-sea mining could wipe away towering spires of minerals built by microbial chemical reactions of deep, ancient fluids. It might destroy species of animals that live at the edges of these hydrothermal vents, ruining habitat for the sea cucumbers, crabs, shrimp, fish, octopus, anemone, and tiny crustaceans that shuffle along our ocean floor, working their way through miles and miles of seafloor.

I do not know the right answer when it comes to the issue of deep-sea mining, but I am certain that the process is not completely safe for ocean life as it currently stands. If safe deep-sea mining practices are indeed developed, maintaining them is going to require careful regulation, oversight, and enforcement.

---

* They are now called Deep-Sea Mining Finance Limited, and some members of this company are now involved in The Metals Company, conducting exploration in the Clarion-Clipperton Zone.

CONCLUSION

THE FUTURE, MAYBE

SINCE THESE intraterrestrials have told us so much about our past, perhaps they can give us a hint about our future as well. Let's think about what the world will be like a thousand years from now. First, we know exactly what those aeonophiles will be doing: very little. A thousand years from now, most of the same individual living cells will be sitting right where we left them, doing exactly what they're doing today. It's refreshing to know that their lives can be so stable when ours are plagued by constant change.

A thousand years from now, I predict that we will have a full list of all the major types of life on Earth—a list that we only gained the technology to start making a few decades ago, thanks to the first DNA revolution. In a thousand years, we will have cataloged all of life's forms, functions, and evolutionary histories. We will have filled in all the gaps between hypothesized chemical reactions that support life and the life that uses them. We will know how much living biomass is produced through chemolithoautotrophy, and we will understand the degree to which all of life on Earth depends on it. People earning college degrees in earth sciences will be required to take biology classes because it will be common knowledge that biology is key to understanding geology. War will be obsolete because we will have realized that its only thermodynamic purpose was to drive innovation through the disruption of stable dissipative structures, which we can do perfectly well through less violent means. Maybe that last one is a stretch, but a girl can dream.

If deep-sea mining ramps up, it might leave a scar visible as a shift in the layering of microbes in the marine sediments. If it develops in the limelight of public awareness, maybe humanity will be able to avoid the worst possible outcomes. We might look to non-equilibrium thermodynamics to suggest what the future holds. One of the hallmarks of life is its ability to spread entropy production out over longer timescales, making processes sustainable over the long term. Ruining an entire habitat to mine metals that are on a one-way trip to a garbage dump is not life's way of doing things. Such a scenario has more of an abiotic feel—like fire burning everything to oblivion. Life seems more likely to develop creative solutions to finding metals, so we can keep that entropy flowing over the long term.

Currently, some carbon-capture techniques work best at high pH, so perhaps newly discovered intraterrestrials adapted to life at high pH will have given us an enzyme that we can use to pull carbon dioxide out of the air more efficiently. Maybe we will have learned enough about subsurface life to know which sorts of geological reservoirs are best for storing carbon and used these new technologies to convert many tons of carbon into a form that is easily sequestered in subsurface reservoirs, stabilizing the climate.

When I imagine the version of me that will be alive a thousand years in the future, I imagine that she'll be writing a book to share her knowledge about *extra*terrestrials with the people of Earth. She will explain how we are finally starting to get a handle on these aliens, describing the planetary bodies where we're finding them and the similarities and differences between their lives and our own.

Today, the surfaces of most of our solar system's planets and moons seem pretty life-free, so in my opinion the most plausible location for extraterrestrial life is beneath their surfaces. Mars, for instance, occasionally releases whiffs of methane into its atmosphere. It is not impossible that a complex ecosystem of living organisms is metabolizing, growing, and dying right now underneath Mars's surface, and that these transient puffs of methane are the only evidence we've seen of them so far.

She might also be writing about subsurface life on Enceladus, which is a moon of Saturn, or Europa, which is a moon of Jupiter. Both of these moons are covered in ice, but underneath this ice are oceans of liquid water that seem like adequate places for life to hide out.[1] It might seem that life on a planetoid completely encased in ice would be very different from life on Earth. But Earth hasn't always been the pretty blue-green ball that it is today. Earth has also spent millions of years completely encapsulated in ice, looking like the spitting image of modern Europa. In fact, these Snowball Earth events have happened multiple times,[2] and life survived them all. Moreover, the intraterrestrials have made it abundantly clear that life doesn't need direct contact with sunlight to exist. There could be alien life sitting happily under the oceans of these moons in the same way that intraterrestrials are sitting inside Earth right now.

This hypothetical future scientist will also write about what it's like to do the "fieldwork" required to learn about these extraterrestrials. But for her, fieldwork will require living on Europa for a few years before returning to Earth. She will describe what it feels like to stand on the icy surface of Europa and see Jupiter looming like a giant orb obstructing nearly the entire horizon. Or how tired she gets when, in her excitement for fieldwork, she forgets to take more than one sleep over the course of a day that lasts about 3.5 Earth days.

Maybe the water that comes roaring to the surface of Europa every time the scientist and her colleagues puncture the ice will carry with it abundant living organisms. This life may include many different kinds of single-celled aliens, but perhaps also little animal-like aliens too. I presume that this life will have cells because I think that life needs an inside and an outside, but I could be wrong about that. These cells may or may not contain DNA or RNA. If they don't, the scientist will have to use new technologies developed in laboratories back on Earth to detect and describe the informational molecules used by the Europans. If she and her colleagues find that Europan life uses a fundamentally different information molecule than DNA and RNA, this will be the first evidence of multiple origins of life in the universe.

She may discover that this alien life is related to Earth life, not through any particular biochemical reaction but via the thermodynamic properties of systems far out of equilibrium. This knowledge will make humanity even more confident that life must exist on (or *in*) the millions of other planetary bodies orbiting other stars in the universe. I mean, if life originated twice in our one little solar system, there must be many other solar systems where this happened as well. By comparing Europan life to Earth life, she'll be able to further refine the thermodynamic boundaries of life. She and her colleagues may discover that Earth's intraterrestrials, which will have been well-known to humans for about a thousand years by then, have been giving them a limited view of what is possible for life. Europan life may show these future scientists that there are ways that life can exist that we cannot even dream up, even with the expanded knowledge that the intraterrestrials have given us.

Teamwork Makes the Dream Work

We are likely to encounter more than a few obstacles as we journey from the present day to that distant future. A few weeks ago, I was in the remote Puna region of Argentina again, gasping for oxygen while sampling hot springs at over four thousand meters above sea level. To reach one of our sites, we needed to cross a rarely traveled high mountain pass, but the road was closed due to landslides. In a nearby village, however, we heard a rumor that someone had made it through with a two-wheel-drive car. So we decided to give it a go, figuring that we had at least as good a shot of success with our four-wheel-drive trucks.

On the gorgeous drive, we passed an elevation of 4,895 meters (16,059 feet) before reaching the hot spring we were seeking. The jagged mountains framed our sinuous single-track dirt path, and I was glad not to be driving, as I was free to stare at the surprisingly green slopes as we moved through them. These were the only mountains we had seen on the trip that were not volcanoes. They were the manifestation of something even more powerful than a volcano—the compression of a continental plate by an oceanic plate subducting beneath it thousands of kilometers away. The resulting towering mountains didn't appear to have any human influences, except for a couple of llama herders on foot.

We had no trouble finding the spring because it sat atop a giant mound of carbonate and iron oxides that had precipitated from the spring's minerals. We happily sampled the spring water, knowing it would be a nice pristine sample with little human interference. Afterward, we munched on an amazing meat pie that my colleague had procured in the small village where we'd stayed the night before, while gazing at the stark unpeopled landscape and its piercing blue sky.

We continued on, eager to leave the mountains and travel to the next site. But a few hundred meters later, we discovered the reason the road had been closed. It was completely blocked by rubble from a recent flash flood. Then, without any planning, hesitation, or pep talks, all eleven of us hopped out of our trucks and started moving rocks. Some of my colleagues studied the road topography to design a navigable path for our tires. Every time someone said, "Pile rocks in front of this boulder here," or, "Fill that gulley there," a flood of rocks would come flying in. We were in constant motion, like a colony of ants, wordlessly joining in on each other's efforts. Once the path looked smooth enough, we moved back to watch the first colleague attempt it. Ugly wrenching noises came out from the bottom of the truck as it scraped in one place, but we made it through! We rushed in to smooth the place where it scraped for the next truck. In the end, all three vehicles survived without any busted tires or brake lines. We had successfully rebuilt the road, using nothing but our own muscles, brains, and literally tons of rocks. When I jumped back in my truck, exhausted, and covered in rock dust and sweat, I felt powerful, like my team and I were capable of anything. We revved the engines and flew across the beautiful mountain landscape.

But our celebration was short-lived. Five minutes later, we screeched to a halt to avoid careening into an entirely new road washout. But this time, we were rock-rolling professionals, functioning like a single organism rather than individual people. We did the work, hopped back in our trucks, and, as you may have guessed, nearly immediately hit another road washout. No sooner had we fixed that one than we hit another one. I was spending so much time in the dust, rolling around with the boulders that I felt like I was just another rock in the landscape. But our magical group strength continued to work.

Our fifth road washout was different. A flashflood had cut a meter-deep trench through the road. We launched a drone to do reconnaissance on the road ahead. It looked solid. If we could just get past this last washout, we'd be home free. But fixing this road would require displacing a big chunk of the hill to even out the grade, which would require an excavator. I marched to the hill with the confidence of an Arctic tern and started pushing on it, unwilling to accept defeat,* but it was clear: the road had won.

We turned around and started the laborious process of retracing our steps. At least our newly reconstructed roads held up as we drove back over them. It would have made sense for us to feel demoralized—we only sampled one site, we worked our tails off, and in the end, the road beat us. But as one of the graduate students later said, this was the best day on the trip. It was the perfect illustration of what will drive this whole enterprise forward, both the fieldwork and the science: people working together toward a common goal. Boltzmann never would have invented modern thermodynamics without admiring Darwin from afar. I would have ignored my DNA sequences from a potentially new domain of life if I hadn't seen the perspectives of my colleagues abroad. Many of us would be ignorant of the growing threat of deep-sea mining if not for our colleagues studying its effects. Discovering the immensity of life inside Earth's crust is a big deal. Realizing its power is going to be a group effort.

I now have the answer to the question that led me to my first submersible dive on the *Johnson-Sea-Link II*: *Are there life-forms hiding inside Earth that are so strange that they change our*

---

* I honestly thought I could move that chunk of mountain until Gerdhard Jessen, my colleague from the Universidad Austral de Chile, looked at me, giggling, and was like, "Uhhhhhhhhhhh, Karen?"

conception of life itself? The answer is an emphatic, "Yes!" We now know that there is a parallel world underneath our feet that is mostly unexplored because it's hard to get into. We need to be careful not to subconsciously twist the saying "Seeing is believing" into "Believing is seeing." When it comes to the intraterrestrials, we need to believe in things that we can't easily see. The intraterrestrials show us that the thermodynamic rules governing life are not nearly as limiting as we might think. The range of chemistry, power, and timescales available for life are vast compared to the limited ones we frequently encounter in our surface world.

This world has been sitting here underneath us the whole time, and we've missed it, as though we've been tricked by a magician's sleight of hand. Nevertheless, the subsurface biosphere is vital to our planet. It holds clues to how life and Earth have co-evolved and how life can exist elsewhere in our solar system.

Putting what we've learned from the intraterrestrials into practice for the betterment of society is going to require efforts from more than just scientists. If scientists can connect with educators, policymakers, engineers, social scientists, entrepreneurs, and many other professions, then maybe, just maybe, we'll make some grand leaps together. Should we put in the hard work to figure out how these intraterrestrials change our conceptions of life and offer solutions to human problems? I'll give the same response I give when someone suggests another cool science adventure, "Why not, let's go!"

ACKNOWLEDGMENTS

MANY EDUCATORS have influenced me over the years, and I have mentioned many of them in this book. Three others emerged early on in my life and set me on the path of science.

My eighth-grade science teacher at Beaufort Middle School, Margery Misenheimer, taught me about what fuels scientific discovery. She revealed to me that "science" was not about dusty textbook facts—it was about the pursuit of the unknown. One day, she came to class with a raw egg in her hand and told us she was going to drop it without breaking it. We'd seen her pull off other impressive experiments, so we craned our necks to observe what would happen this time. She held the egg high in front of her, dropped it, and the egg went "splat" all over the linoleum. She looked up at us, smiled, and said, "Did you see?! It worked!" We were like, "Uhhhhh, no it didn't . . . ," to which she replied, "Oh, sure, it broke when it hit the floor, but while it was dropping through the air it was fine." This messy little prank stuck with me because it made me understand that she was going out on a limb to urge us to think critically and carefully. She was warning us not to accept things at face value; a little bit of creative analysis sometimes reveals what you might be missing.

After my sophomore year in high school, I attended Summer Ventures in Science and Mathematics, a publicly funded summer camp at North Carolina Central University. For me, the public funding was essential. My family did not have the resources to send me to a fancy sleepaway camp to learn science from real scientists; Summer Ventures would provide me with just such an opportunity. I knew how lucky I was to get to be a part of this camp, and I tried to soak up as much as I could. I chose a course in microbiology taught by John Mayfield, an expert in fungi that degrade fallen trees in the forests. What he taught us over that sweltering summer ended up determining the course of my future career, and, really, my whole life. He told us that microbes, which are tiny single-celled bacteria, archaea, and eukaryotes, are the hidden forces behind almost everything that happens on Earth. This was the first time I heard such an assertion, and I haven't stopped thinking about it since.

Kathleen Howard, professor of chemistry at Swarthmore College, helped me transition into a professional scientist. She was my advisor for my undergraduate thesis research at Swarthmore College. Besides teaching me the finer details of nuclear magnetic resonance spectroscopy and lipid membrane biophysical chemistry, she once said to me, "In research, there are more failures than successes. If your project fails Monday through Thursday, but finally works on Friday, and you can still go home for the weekend satisfied with how you spent your week, then research is the job for you. But if you're going to be frustrated that you wasted the first four days of the week, then you should go do something else for a career." These wise words have stuck with me through my many failures over the years, and I am forever grateful to her for opening the doors to research for me.

I also want to acknowledge three dear and influential scientists who were instrumental in the work of advancing the field of deep subsurface biology and have also been of great importance to me as a scientist. My perspectives on thermodynamics and life were deepened by the work of Jan P. Amend, professor of earth sciences and dean at the University of Southern

California, who passed away in 2024. Jan had a wonderful way of diffusing any tense scientific discussion with a witty remark and an infectious giggle that was like medicine for the soul. He also directly contributed to this book by reading over early drafts of the manuscript. Tullis C. Onstott, professor of earth sciences at Princeton University, passed away in 2021. He was a force for goodness in this world, and I owe a lot to scientists like Tullis who were welcoming and respectful of younger scientists such as myself. I would also like to acknowledge Katrina J. Edwards, professor of earth sciences at the University of Southern California, who passed away in 2014. Her light burned so bright that she ignited an entire field of study.

I am deeply indebted to my husband and colleague, Drew Steen, who helped me get my thinking straight on the most difficult parts of this book, reminded me that too many words is the enemy, and gave me the kind of love and support that I'm lucky to have. Our children, Clara and Mary Jon Steen, encouraged me, listened to me read aloud, gave me great feedback, and put up with me "clicky-clacking" on my computer. My mother and stepfather, Pennylloyd and John Baldridge, my husband's parents, Bob and Ruth Steen, and my aunt and uncle Nancy Ferguson and Tom Orum have not just been sources of moral support and cheerleading but have also slogged through many drafts of this book themselves. My father, Frank Lloyd, would have been over the moon to see that I wrote a book. He passed away in 2014, long after making sure I had a childhood full of adventure around every corner.

I would also like to thank the Biology Meets Subduction crew, primarily Peter Barry, Agostina Chiodi, Maarten de Moor, Donato Giovannelli, Gerdhard Jessen, Jacopo Pasotti, and Carlos Ramírez for listening to me talk about my book while adventuring through the Andes and elsewhere, and for not complaining too much about me balancing scientific gear with my feet so I could write while pumping fluids from hot springs.

I am deeply indebted to Alison Kalett at Princeton University Press, who talked me into writing a book in the first place; Hallie Schaefer, also at Princeton University Press, who helped edit; Dana Henricks, who copyedited the manuscript; and the folks at Moon & Company (Amanda Moon, Thomas LeBien, and James Brandt), who have read this book as many times as I have and helped me whip it into shape.

The following people also read versions of this book, gave me helpful edits, advice on book writing, and photos from the field, and just generally rocked my world: Andrey Abramov, Chukwufumnanya Abuah, Jennifer Baily, Jillian Banfield, James Bradley, Joy Buongiorno, Gage Coon, Mark Faller, Danik Forsman, Avery Fulford, Robert Hazen, Julie Huber, Brian Hynek, Oliver Jeffers, Bo Barker Jørgensen, Ellen Lalk, Douglas E. LaRowe, Finley Lloyd, Jeffrey Marlow, Howard Mendlovitz, Flavia Migliaccio, Sarina Mitchell, Francesco Montemagno, Beth Orcutt, Dorthe Petersen, Catherine Pratt, Lars Schreiber, Katie Sipes, Douglas Sofer, Ramunas Stepanauskas, Andreas Teske, Elizabeth Trembath-Reichert, Steve Turner, Joseph Vallino, Tatiana Vishnivetskaya, Leketha Williams, and everyone at Ferguson Camp at Elfin Lake and St. James Episcopal Church in Knoxville, Tennessee. Without the enthusiasm and encouragement of all of these people, I never would have seen this thing through to the end.

REFERENCES

Introduction

1. Pedersen, K. Exploration of deep intraterrestrial microbial life: Current perspectives. *FEMS Microbiol. Lett.* **185**, 9–16 (2000).

2. Onstott, T. C. *Deep Life: The Hunt for the Hidden Biology of Earth, Mars, and Beyond* (Princeton University Press, 2017).

3. Valentine, D. L. Microbiology: Intraterrestrial lifestyles. *Nature* **496**, 176–77 (2013).

1. Is There a "Habitat" inside Earth's Crust?

1. Chivian, D., et al. Environmental genomics reveals a single-species ecosystem deep within Earth. *Science* **322**, 275–78 (2008).

2. Ruff, S. E., et al. Hydrogen and dark oxygen drive microbial productivity in diverse groundwater ecosystems. *Nat. Commun.* **14**, 3194 (2023); Kraft, B., et al. Oxygen and nitrogen production by an ammonia-oxidizing archaeon. *Science* **375**, 97–100 (2022).

3. LaRowe, D. E., Burwicz, E., Arndt, S., Dale, A. W., & Amend, J. P. Temperature and volume of global marine sediments. *Geology* **45**, 275–78 (2017).

4. Karner, M. B., DeLong, E. F., & Karl, D. M. Archaeal dominance in the mesopelagic zone of the Pacific Ocean. *Nature* **409**, 507–10 (2001); Parkes, R. J., Cragg, B. A., & Wellsbury, P. Recent studies on bacterial populations and processes in subseafloor sediments: A review. *Hydrogeol. J.* **8**, 11–28 (2000).

5. Kallmeyer, J., Pockalny, R., Adhikari, R. R., Smith, D. C., & D'Hondt, S. Global distribution of microbial abundance and biomass in subseafloor sediment. *Proc. Natl. Acad. Sci. USA* **109**, 16213–16 (2012).

6. Magnabosco, C., et al. The biomass and biodiversity of the continental subsurface. *Nat. Geosci.* **11**, 707–17 (2018).

7. Lloyd, K. G., et al. Spatial structure and activity of sedimentary microbial communities underlying a *Beggiatoa* spp. mat in a Gulf of Mexico hydrocarbon seep. *PLOS One* **5**, e8738 (2010).

2. Cracking into Solid Earth

1. Heuer, V. B., et al. Temperature limits to deep subseafloor life in the Nankai Trough subduction zone. *Science* **370**, 1230–34 (2020).

2. Onstott, T. C. *Deep Life: The Hunt for the Hidden Biology of Earth, Mars, and Beyond* (Princeton University Press, 2017).

3. Summit, M., & Baross, J. A. A novel microbial habitat in the mid-ocean ridge subseafloor. *Proc. Natl. Acad. Sci.* **98**, 2158–63 (2001).

3. The Two DNA Revolutions

1. Rappé, M. S., & Giovannoni, S. J. The uncultured microbial majority. *Annu. Rev. Microbiol.* **57**, 369–94 (2003).

2. Rappé, M. S., & Giovannoni, S. J. The uncultured microbial majority. *Annu. Rev. Microbiol.* **57**, 369–94 (2003); Teske, A., & Sørensen, K. B. Uncultured archaea in deep marine subsurface sediments: Have we caught them all? *ISME J.* **2**, 3–18 (2008).

3. Sogin, M. L., et al. Microbial diversity in the deep sea and the underexplored "rare biosphere." *Proc. Natl. Acad. Sci. USA* **103**, 12115–20 (2006); Huber, J. A., et al. Microbial population structures in the deep marine biosphere. *Science* **318**, 97–100 (2007).

4. Lloyd, K. G., Lapham, L., & Teske, A. An anaerobic methane-oxidizing community of ANME-1b archaea in hypersaline Gulf of Mexico sediments. *Appl. Environ. Microbiol.* **72**, 7218–30 (2006).

5. Louca, S., Mazel, F., Doebeli, M., & Parfrey, L. W. A census-based estimate of Earth's bacterial and archaeal diversity. *PLOS Biol.* **17**, e3000106 (2019).

4. Humans and Other Plants

1. Margulis, L., & Schwartz, K. V. *Five Kingdoms: An Illustrated Guide to the Phyla of Life on Earth* (Freeman & Company, W. H., 1982).

2. Woese, C. R. Bacterial evolution. *Microbiol. Rev.* **51**, 221–71 (1987).

3. Mayr, E. Perspective: Two empires or three? *Comp. Gen. Pharmacol.* **95**, 9720–23 (1998).

4. Mayr, E. Perspective: Two empires or three? *Comp. Gen. Pharmacol.* **95**, 9720–23 (1998).

5. Baker, B. J., et al. Diversity, ecology and evolution of Archaea. *Nat. Microbiol.* **29**, 887–900 (2020).

6. Hug, L. A., et al. A new view of the tree of life. *Nat. Microbiol.* **1**, 1–6 (2016).

7. Jagus, R., Bachvaroff, T. R., Joshi, B., & Place, A. R. Diversity of eukaryotic translational initiation factor eIF4E in protists. *Comp. Funct. Genomics* **2012**, 1–21 (2012).

8. Jurgens, G., et al. Identification of novel Archaea in bacterioplankton of a boreal forest lake by phylogenetic analysis and fluorescent in situ hybridization. *FEMS Microbiol. Ecol.* **34**, 45–56 (2000).

9. Inagaki, F., et al. Microbial communities associated with geological horizons in coastal subseafloor sediments from the sea of Okhotsk. *Appl. Environ. Microbiol.* **69**, 7224–35 (2003).

10. Takai, K. E. N., Moser, D. P., Flaun, M. D. E., Onstott, T. C., & Fredrickson, J. K. Archaeal diversity in waters from deep South African gold mines. *Appl. Environ. Microbiol.* **67**, 5750–60 (2001).

11. Lloyd, K. G., et al. Predominant archaea in marine sediments degrade detrital proteins. *Nature* **496**, 215–18 (2013).

12. Edwards, K. J. An archaeal iron-oxidizing extreme acidophile important in acid mine drainage. *Science* **287**, 1796–99 (2000).

13. Tyson, G. W., et al. Community structure and metabolism through reconstruction of microbial genomes from the environment. *Nature* **428**, 37–43 (2004).

14. Wrighton, K. C., et al. Fermentation, hydrogen, and sulfur metabolism in multiple uncultivated bacterial phyla. *Science* **337**, 1661–66 (2012).

15. Meng, J., et al. Genetic and functional properties of uncultivated MCG archaea assessed by metagenome and gene expression analyses. *ISME J.* **8**, 650–59 (2014).

5. How to Live inside a Volcano

1. Madshus, I. H. Regulation of intracellular pH in eukaryotic cells. *Biochem. J.* **250**, 1–8 (1988).

2. Golyshina, O. V., Golyshin, P. N., Timmis, K. N., & Ferrer, M. The "pH optimum anomaly" of intracellular enzymes of Ferroplasma acidiphilum. *Environ. Microbiol.* **8**, 416–25 (2006).

3. Baker-Austin, C., & Dopson, M. Life in acid: pH homeostasis in acidophiles. *Trends Microbiol.* **15**, 165–71 (2007).

4. Hynek, B. M., Rogers, K. L., Antunovich, M., Avard, G., & Alvarado, G. E. Lack of microbial diversity in an extreme Mars analog setting: Poás volcano, Costa Rica. *Astrobiology* **18**, 923–33 (2018).

5. Russell, M. J., Hall, A. J., & Martin, W. Serpentinization as a source of energy at the origin of life. *Geobiology* **8**, 355–71 (2010).

6. Krulwich, T. A., et al. Adaptive Mechanisms of Extreme Alkaliphiles. In *Extremophiles Handbook* (ed. Horikoshi, K.), 119–39 (Springer Japan, 2011). https://doi:10.1007/978-4-431-53898-1_7.

7. Suzuki, S., et al. Unusual metabolic diversity of hyperalkaliphilic microbial communities associated with subterranean serpentinization at the Cedars. *ISME J.* **11**, 2584–98 (2017).

6. Breathing Rocks

1. Larowe, D. E., & Cappellen, P. V. Degradation of natural organic matter: A thermodynamic analysis. *Geochim. Cosmochim. Acta* **75**, 2030–42 (2011).

2. Nealson, K. H. Sediment bacteria: Who's there, what are they doing, and what's new? *Annu. Rev. Earth Planet. Sci.* **25**, 403–34 (1997).

3. Lloyd, J. R. Microbial reduction of metals and raionuclides. *FEMS Microbiol. Rev.* **27**, 411–25 (2003).

4. Lovley, D. R., & Malvankar, N. S. Minireview: Seeing is believing: Novel imaging techniques help clarify microbial nanowire structure and function. *Environ. Microbiol.* **17**, 2209–15 (2015).

5. Kuenen, J. G. Anammox bacteria: From discovery to application. *Nat. Rev. Microbiol.* **6**, 320–26 (2008).

6. Lu, Guang-Sin, LaRowe, D. E., & Amend, J. P. Bioenergetic potentials in terrestrial, shallow-sea and deep-sea hydrothermal systems. *Chem. Geol.* **583**, 120449 (2021).

7. Living on the Edge

1. Buongiorno, J., et al. Complex microbial communities drive iron and sulfur cycling in arctic fjord sediments. *Appl. Environ. Microbiol.* **85**, 1–16 (2019).

2. Wehrmann, L. M., et al. Iron and manganese speciation and cycling in glacially influenced high-latitude fjord sediments (West Spitsbergen, Svalbard): Evidence for a benthic recycling-transport mechanism. *Geochim. Cosmochim. Acta* **141**, 628–55 (2014).

3. Zinder, S. H. Syntrophic acetate oxidation and "reversible acetogenesis." In *Acetogenesis* (ed. Drake, H. L.), 386–415 (Springer US, 1994). https://doi:10.1007 /978-1-4615-1777-1_14.

4. Ferry, J. G. Acetate metabolism in anaerobes from the domain Archaea. *Life Basel Switz.* **5**, 1454–71 (2015).

5. Lloyd, K. G., Alperin, M. J., & Teske, A. Environmental evidence for net methane production and oxidation in putative ANaerobic MEthanotrophic (ANME) archaea. *Environ. Microbiol.* **13**, 2548–64 (2011).

6. LaRowe, D. E., & Amend, J. P. The energetics of fermentation in natural settings. *Geomicrobiol. J.* **36**, 492–505 (2019).

7. Bradley, J. A., et al. Widespread energy limitation to life in global subseafloor sediments. *Sci. Adv.* **6**, eaba0697 (2020).

8. Immortal Microbes

1. Lloyd, K. G., et al. Effects of dissolved sulfide, pH, and temperature on growth and survival of marine hyperthermophilic archaea. *Appl. Environ. Microbiol.* **71**, 6383–87 (2005).

2. Lennon, J. T., & Jones, S. E. Microbial seed banks: The ecological and evolutionary implications of dormancy. *Nat. Rev. Microbiol.* **9**, 119–30 (2011).

3. Alnimr, A. M. Dormancy models for mycobacterium tuberculosis: A minireview. *Braz. J. Microbiol.* **46**, 641–47 (2015).

4. Davis, K. E. R., Joseph, S. J., & Janssen, P. H. Effects of growth medium, inoculum size, and incubation time on culturability and isolation of soil bacteria. *Appl. Environ. Microbiol.* **71**, 826–34 (2005).

5. Könneke, M., et al. Isolation of an autotrophic ammonia-oxidizing marine archaeon. *Nature* **437**, 543–46 (2005); Rappe, M. S., Connon, S. A., Vergin, K. L., & Giovannoni, S. J. Cultivation of the ubiquitous SAR11 marine bacterioplankton clade. *Nature* **418**, 630–33 (2002).

6. Nauhaus, K., Albrecht, M., Elvert, M., Boetius, A., & Widdel, F. In vitro cell growth of marine archaeal-bacterial consortia during anaerobic oxidation of methane with sulfate. *Environ. Microbiol.* **9**, 187–96 (2007); Imachi, H. et al. Isolation of an archaeon at the prokaryote–eukaryote interface. *Nature* **577**, 519–25 (2020); Katayama, T., et al. Isolation of a member of the candidate phylum "Atribacteria" reveals a unique cell membrane structure. *Nat. Commun.* **11**, 1–9 (2020). https://doi:10.1038/s41467-020-20149-5.

7. Bradley, J. A., et al. Widespread energy limitation to life in global subseafloor sediments. *Sci. Adv.* **6**, eaba0697 (2020); Hoehler, T. M., & Jørgensen, B. B. Microbial life under extreme energy limitation. *Nat. Rev. Microbiol.* **11**, 83–94 (2013); Larowe, D. E., & Amend, J. P. Power limits for microbial life. *Front. Microbiol.* **6**, 1–11 (2015).

8. Lloyd, K. G., et al. Evidence for a growth zone for deep subsurface microbial clades in near-surface anoxic sediments. *Appl. Environ. Microbiol.* **86**, 1–13 (2020); Starnawski, P., et al. Microbial community assembly and evolution in subseafloor sediment. *Proc. Natl. Acad. Sci.* **114**, 2940–45 (2017); Braun, S., et al. Size and carbon content of sub-seafloor microbial cells at Landsort Deep, Baltic Sea. *Front. Microbiol.* **7**, 1–13 (2016).

9. Trembath-Reichert, E., et al. Methyl-compound use and slow growth characterize microbial life in 2-km-deep subseafloor coal and shale beds. *Proc. Natl. Acad. Sci. USA.* **114**, E9206–E9215 (2017).

10. Braun, S., et al. Size and carbon content of sub-seafloor microbial cells at Landsort Deep, Baltic Sea. *Front. Microbiol.* **7**, 1–13 (2016).

11. Starnawski, P., et al. Microbial community assembly and evolution in subseafloor sediment. *Proc. Natl. Acad. Sci.* **114**, 2940–45 (2017); Walsh, E. A., et al. Bacterial diversity and community composition from seasurface to subseafloor. *ISME J.* **10**, 979–89 (2016).

12. Teske, A., & Sørensen, K. B. Uncultured archaea in deep marine subsurface sediments: Have we caught them all? *ISME J.* **2**, 3–18 (2008); Durbin, A. M., & Teske, A. Archaea in organic-lean and organic-rich marine subsurface sediments: An

environmental gradient reflected in distinct phylogenetic lineages. *Front. Microbiol.* **3**, 168 (2012).

13. Steen, A. D., et al. Kinetics and identities of extracellular peptidases in subsurface sediments of the White Oak River Estuary, NC. *Appl. Environ. Microbiol.* **85**, 1–14 (2019).

14. Bird, J. T., et al. Uncultured microbial phyla suggest mechanisms for multithousand-year subsistence in Baltic Sea sediments. *mBio* **10**, 1–15 (2019).

15. Finkel, S. E. Long-term survival during stationary phase: Evolution and the GASP phenotype. *Nat. Rev. Microbiol.* **4**, 113–20 (2006).

16. Lloyd, K. G., et al. Evidence for a growth zone for deep subsurface microbial clades in near-surface anoxic sediments. *Appl. Environ. Microbiol.* **86**, 1–13 (2020).

17. Steen, A. D., & Arnosti, C. Long lifetimes of β-glucosidase, leucine aminopeptidase, and phosphatase in Arctic seawater. *Mar. Chem.* **123**, 127–32 (2011).

18. Pianka, E. R. On r- and K-Selection. *Am. Soc. Nat.* **104**, 592–97 (1970).

9. Rethinking Our Beginnings

1. Vernadsky, V. I. *The Biosphere* (Copernicus, 1998).

2. Knittel, K., Lo, T., Boetius, A., Kort, R., & Amann, R. Diversity and distribution of methanotrophic archaea at cold seeps. *Appl. Environ. Microbiol.* **71**, 467–79 (2005).

3. Spang, A., et al. Complex archaea that bridge the gap between prokaryotes and eukaryotes. *Nature* **521**, 173–79 (2015). https://doi:10.1038/nature14447.

4. Imachi, H., et al. Isolation of an archaeon at the prokaryote–eukaryote interface. *Nature* **577**, 519–25 (2020).

10. Equilibrium Is Death

1. Schneider, E. D., & Sagan, D. *Into the Cool: Energy Flow, Thermodynamics, and Life* (University of Chicago Press, 2005).

2. Prigogine, I., & Stengers, I. *Order out of Chaos: Man's New Dialogue with Nature* (Bantam, 1984).

3. Vallino, J. J. Ecosystem biogeochemistry considered as a distributed metabolic network ordered by maximum entropy production. *Philos. Trans. R. Soc. B Biol. Sci.* **365**, 1417–27 (2010).

11. What Can Intraterrestrials Do for Us?

1. Liang, R., et al. Predominance of anaerobic, spore-forming bacteria in metabolically active microbial communities from ancient Siberian permafrost. *Appl. Environ. Microbiol.* **85**, e00560–19 (2019).

2. Sipes, K., et al. Eight metagenome-assembled genomes provide evidence for microbial adaptation in 20,000- to 1,000,000-year-old Siberian permafrost. *Appl. Environ. Microbiol.* **87**, e00972–21 (2021).

3. Waldrop, M. P., et al. Molecular investigations into a globally important carbon pool: Permafrost-protected carbon in Alaskan soils. *Glob. Change Biol.* **16**, 2543–54 (2010). https://doi:10.1111/j.1365-2486.2009.02141.x.

4. Mann, P. J., et al. Evidence for key enzymatic controls on metabolism of Arctic river organic matter. *Glob. Change Biol.* **20**, 1089–1100 (2014).

5. Panko, Ben. World's largest carbon capture plant opens in Iceland. *Smithsonian Magazine* (2021). https://www.smithsonianmag.com/smart-news/worlds-largest-carbon-capture-plant-opens-iceland-180978620/.

6. Tyne, R. L., et al. Rapid microbial methanogenesis during CO_2 storage in hydrocarbon reservoirs. *Nature* **600**, 670–74 (2021).

7. Mitchell, A. C., Dideriksen, K., Spangler, L. H., Cunningham, A. B., & Gerlach, R. Microbially enhanced carbon capture and storage by mineral-trapping and solubility-trapping. *Environ. Sci. Technol.* **44**, 5270–76 (2010).

8. Cronan, David S. Manganese nodules. In *Encyclopedia of Ocean Sciences (Third Edition)* (ed. Cochran, J. K., Bokuniewicz, H. J., & Yager, P. L.), vol. 5, 607–14 (Academic Press, 2019).

9. International Seabed Authority. About ISA. https://www.isa.org.jm/about-isa/.

10. Hein, J. R., et al. Cobalt-rich ferromanganese crusts in the Pacific. In *Handbook of Marine Mineral Deposits* (ed. Cronan, D. S.), 239–80 (CRC Press, 1999).

11. Dopffel, N., Jansen, S., & Gerritse, J. Microbial side effects of underground hydrogen storage: Knowledge gaps, risks and opportunities for successful implementation. *Int. J. Hydrog. Energy* **46**, 8594–8606 (2021).

Conclusion: The Future, Maybe

1. Hand, K. P. *Alien Oceans: The Search for Life in the Depths of Space* (Princeton University Press, 2020).

2. Hoffman, P. F., et al. Snowball Earth climate dynamics and cryogenian geology-geobiology. *Sci. Adv.* **3**, e1600983 (2017).

INDEX